W9-AIG-353

PREVENTING THE NEXT PANDEMIC

OTHER BOOKS BY PETER J. HOTEZ

Forgotten People, Forgotten Diseases: The Neglected Tropical Diseases and Their Impact on Global Health and Development, second edition, 2013

Blue Marble Health: An Innovative Plan to Fight Diseases of the Poor amid Wealth, 2016

Vaccines Did Not Cause Rachel's Autism: My Journey as a Vaccine Scientist, Pediatrician, and Autism Dad, 2018

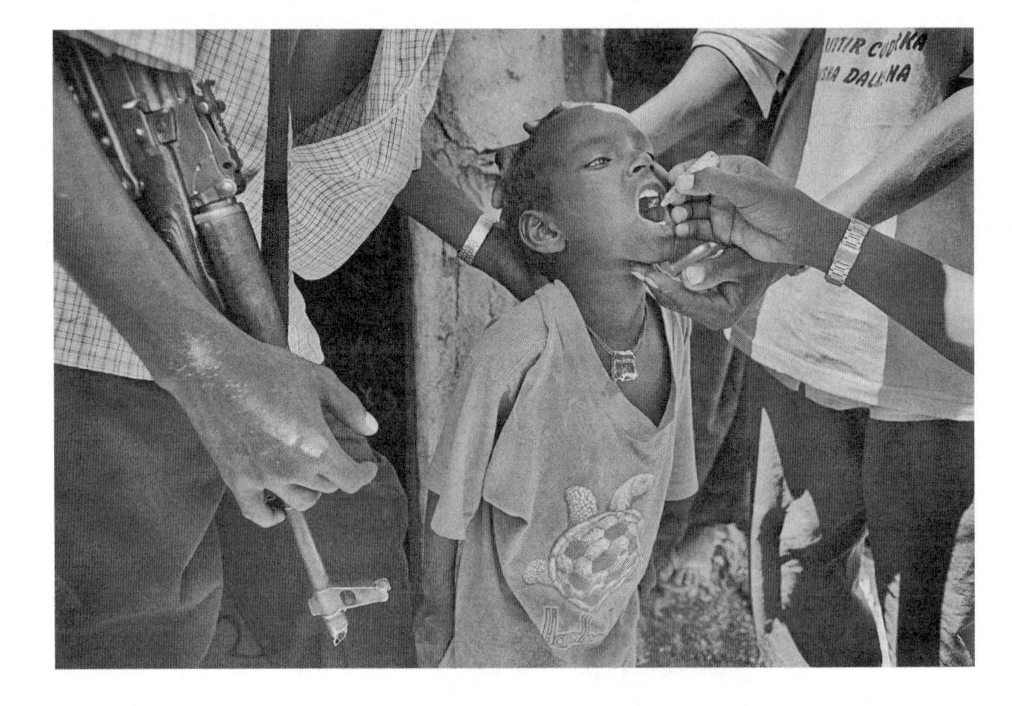

PETER J. HOTEZ, MD, PHD

PREVENTING THE NEXT PANDEMIC

Vaccine Diplomacy
in a Time of Anti-science

Johns Hopkins University Press
Baltimore

Printed in the United States of America on acid-free paper
9 8 7 6 5 4 3 2 1

Johns Hopkins University Press
2715 North Charles Street
Baltimore, Maryland 21218-4363
www.press.jhu.edu

Library of Congress Cataloging-in-Publication Data

Names: Hotez, Peter J., author.
Title: Preventing the next pandemic : vaccine diplomacy in a time of anti-science / Peter J. Hotez, MD, PhD.
Description: Baltimore : Johns Hopkins University Press, [2021] | Includes bibliographical references and index.
Identifiers: LCCN 2020018801 | ISBN 9781421440385 (hardcover ; alk. paper) | ISBN 9781421440392 (ebook)
Subjects: MESH: Pandemics—prevention & control | Vaccines | Vaccine-Preventable Diseases—prevention & control | Diplomacy | International Cooperation | Anti-Vaccination Movement
Classification: LCC RA566.27 | NLM WA 105 | DDC 362.19698—dc23
LC record available at https://lccn.loc.gov/2020018801

A catalog record for this book is available from the British Library.

Frontispiece: Child receiving polio vaccine under armed protection in Somalia. Photograph by Sebastião Salgado.

Johns Hopkins University Press uses environmentally friendly book materials, including recycled text paper that is composed of at least 30 percent post-consumer waste, whenever possible.

To my amazing colleagues and professionals with the United States Department of State, White House Office of Science and Technology Policy, and US embassies abroad, when I served as US science envoy (2015–16) and now as a governor of the US-Israel Binational Science Foundation

Special thank you and dedication to the anchors and producers at CNN, Fox News, and MSNBC for the opportunity and privilege to speak to the nation during the COVID-19 pandemic

Contents

Preface

This book reports on the recent and unexpected rise in infectious and tropical diseases owing to twenty-first-century forces: war and conflict, shifting poverty, urbanization, climate change, and a new, troubling anti-science. I explain how through vaccine diplomacy we address this new world order in disease and global health.

My activities and the people I met as both United States science envoy, a position I held in the Obama administration in 2015 and 2016, and subsequently as a member of the board of governors of the US-Israel Binational Science Foundation in the Trump administration, inspired the book. Infectious and tropical diseases are now abruptly arising in multiple hotspot areas across the globe. They include the Northern Triangle of Central America, Venezuela, and neighboring countries in South America; the conflict zones on the Arabian Peninsula, especially Syria, Iraq, and Yemen; Democratic Republic of Congo, South Sudan, and elsewhere in central and eastern sub-Saharan Africa; and several other places. Now, the planet is being consumed by a COVID-19 pandemic. I explain how vaccine diplomacy might offer new solutions to the devastation caused by infection in these regions and how it might prevent future disease catastrophes. I also report on our current vaccine diplomacy activities and my work as a vaccinologist developing new neglected disease vaccines for the world's poorest people.

Once again, I want to express profound gratitude to my very important mentors and bosses at Baylor College of Medicine—Dr. Paul Klotman—and at Texas Children's Hospital—Mr. Mark Wallace.

I would like to thank my mentors and colleagues in both the US State Department and the White House Office of Science and Technology Policy (OSTP), who provided the oversight for the US Science Envoy Program. They include Catherine Novelli, former undersecretary of state for economic growth, energy, and the environment; assistant secretaries Jonathan Margolis and Judith Garber (now ambassador to Cyprus); John Holdren, former director of the White House OSTP and senior adviser to President Barack Obama; ambassadors Joseph Westphal and Dwight Lamar Bush Sr.; and Douglas Apostle, Kimberly Coleman, Daisy Dix, Donya Eldridge, Kay Hairston, Kia Henry, Patricia Hill, Stephanie Hutchison, Bryce Isham, Mohamed Khalil, Sara Klucking, Amani Meki, Bruce Ruscio, and Matthew West. From the US-Israel Binational Science Foundation, I want to thank my current and past colleagues or fellow board of governors members, including Jared Bank, Cathy Campbell, Howard Cedar, Joshua Gordon, Bracha Halaf, Heni Haring, Andrew Hebbeler, Avi Israeli, Rebecca Lynn Keiser, Yair Rotstein, Ido Sofer, Riju Srimal, and Ishi Talmon.

I also would like to acknowledge my mentors at Rice University's Baker Institute for Public Policy, former ambassador Edward Djerejian and Neal Lane; at the Scowcroft Institute of International Affairs, Texas A&M University, ambassador Andrew Natsios; and at the Hagler Institute for Advanced Study at Texas A&M University, Gregory Brian Colwell, Clifford Fry, John Junkins, Gerald Parker, deans John August, Eleanor Green, Jay Maddock, and vice chancellor, Dr. Carrie Byington. I also thank Baylor University's president Linda Livingstone, provost

Nancy Brickhouse, dean Lee Nordt, Lori Baker, Richard Sanker, Dwayne Simmons, and former Baylor president, Judge Kenneth Starr, and his wife, Alice Mendell Starr, for their support.

I also want to thank Dr. David Kaslow, Vice President of Essential Medicines at PATH, and his outstanding team of scientists, including Dr. Fred Cassels, Deborah Higgins, and many others, for their advice in advancing our new COVID-19 vaccine.

Once again, I am grateful to Nathaniel Wolf for his stalwart support and long-standing editorial assistance and guidance, to Douglas Osejo Soriano for administrative assistance, and to Dr. Maria Elena Bottazzi and the vaccine research team at the Texas Children's Hospital Center for Vaccine Development for their support and advice. I also want to thank Robin Coleman and my publishers at Johns Hopkins University Press, who again rolled the dice with me for this latest book. Finally, I am deeply grateful to my wife, Ann Hotez, and my four adult children, Matthew, Emily, Rachel, and Daniel, and their spouses or significant others, Brooke Hotez, Yan Slavinskiy, and Alexandra Pfeiffer.

PREVENTING THE NEXT PANDEMIC

1:

A New Post-2015 Urgency

It was my first *dewaniya*, a gathering of men (or usually men) who meet to discuss business or important social issues of the day. Dewaniyas are a mainstay of the political and social life of the Middle East, and this evening I was a guest of honor in Dammam, the major city of Saudi Arabia's Eastern Province and the epicenter of its oil fields and industry. The food was excellent, as it usually is in the Kingdom of Saudi Arabia, and accompanied by lots of strong tea and a type of Arabic coffee with cardamom served in the Bedouin tradition. I especially liked the dates and other types of candied fruit.

The Saudi people are terrific hosts and go out of their way to make their guests feel special and welcome. But I also sensed some unspoken tension, and I had an important mission beyond enjoying the dates and coffee. The year was 2015, and I was a few months into my first one-year term as US science envoy. This was a new position created jointly between the State Department and the White House Office of Science and Technology Policy. It was established as part of America's new outreach to the Muslim world, beginning when President Barack Obama traveled to Cairo, Egypt, in his first year in office. Secretary of State Hillary Rodham Clinton then shaped the position to its current form. Both leaders recognized the power of American science to complement and enhance US foreign policy.

At the end of 2014, I was invited to serve as US science envoy for the Middle East and North Africa. Presumably, I was selected because of my expertise in the science of vaccinology and vaccine development, together with my interests in the neglected and emerging diseases of the region. By then, through my writings, I had also become known for promoting the concept of "vaccine diplomacy," roughly referring to simultaneous scientific and diplomatic opportunities between nations, with an overriding objective to jointly develop and test vaccines as a means to promote health, security, and peace. My desire to incorporate vaccine diplomacy into US foreign policy was inspired by the achievements of the late Dr. Albert Sabin, who developed the oral polio vaccine jointly with Soviet scientists at the height of the Cold War during the late 1950s and 1960s.

As US science envoy, my priority was the Kingdom of Saudi Arabia, which grew out of a unique confluence of overlapping medical urgencies and the geopolitics of the region. By 2015, the Kingdom found itself situated between two major conflict zones on the Arabian Peninsula—Iraq and Syria to the north and Yemen to the south. Because of war and occupation, health systems had collapsed in these areas, resulting in the cessation of vaccination initiatives and of programs to control insect vectors that spread diseases such as dengue, Rift Valley fever, and leishmaniasis. As a result, the Arabian Peninsula had become a new hot zone of infectious and tropical diseases. Higher temperatures and shifting rainfall patterns were also contributing to the rise in disease. The Saudis were concerned about the risk of disease spreading into their homeland, and I was there to explore our joint development of vaccines and other countermeasures, possibly with either my vaccine laboratory or others in the United States.

Complicating this situation was the problem of Iran, the center of Shia Islam and a bitter rival of Saudi Arabia, the dominant

Sunni Muslim nation in the region. These tensions had helped to fuel conflict on the Arabian Peninsula, especially in Yemen. Iran also promoted unrest in Saudi's Eastern Province, where a Shia minority population lived. My dewaniya happened just a few months after a Shia shrine in eastern Saudi Arabia was attacked by masked gunmen. More important, the Saudis were also widely aware that the Obama administration had made a recent geopolitical pivot toward Iran and was negotiating a nuclear disarmament and peace treaty with that country's leadership. Many Saudis viewed this act as a historic betrayal of their special relationship with the United States. Then, to top it all off, I was in Dammam following Donald J. Trump's announcement of his first presidential campaign, when he began making inflammatory statements about Muslims. By the end of 2015, he had called for a total ban on all Muslims entering the United States.

As one might imagine, the combination of America's high-level discussions with Iran, the Trump candidacy, and a recent attack on a Shia shrine might make this not the most propitious time to be drinking Arabic coffee in the Eastern Province of Saudi Arabia. Perhaps it was also not a great time to offer vaccine diplomacy as a meaningful response to heightened tensions between our two nations. On the other hand, I was incredibly energized and excited. After all, when would I get another opportunity like this to embark on a vaccine diplomatic mission? Certainly, I thought, US-Saudi relations were not nearly as challenging as US-Soviet relations were in the years immediately following the Sputnik launch in 1957, when Dr. Sabin began his partnership to develop the polio vaccine.

For me, a watershed moment in my diplomatic visits occurred when a senior Saudi official came up to me in a reassuring way. He spoke to me specifically about the Trump Muslim ban and said something more or less along the following lines: "Peter, we're not too worried. We realize these statements are nonsense. Look, I did my PhD at Iowa State University. For years, I lived with a family in Ames, Iowa. I loved every minute of it. I know how the real Americans think; everything will be fine." It was then I realized how America's research universities and institutes rank among our greatest national treasures and provide an important and promising face for our nation. I gave him a big hug—in part, because his remarks broke the tensions and I was relieved—but also because it reinforced my conviction that our scientists and science prowess represent the best of America and the side we must present to the world.

I made several visits to Saudi Arabia in 2015 and 2016, which ultimately resulted in an important and new vaccine collaboration between our laboratory at Texas Children's Hospital and Baylor College of Medicine and King Saud University. We are exploring additional collaborations, possibly including the King Abdullah University of Science and Technology, considered the Saudi version of MIT. However, Saudi Arabia is just one nation, and there were many others where increasingly I saw diseases return through war or political instability, or nations where diseases are now returning because of other twenty-first-century forces. I benchmark 2015 as the year when infectious and tropical diseases began this new ascent. This book reports on how war, political instability, climate change, and other determinants such as anti-science movements combine in unique and interesting ways to introduce or bring back disease. It also describes how vaccine diplomacy might offer new and innovative solutions.

An Unexpected Shift

Just a few years ago, many of us in the global health policy community were thrilled at the prospect of eliminating catastrophic infectious and tropical diseases. The year 2015 was approaching, and we would soon complete one of the most ambitious global development initiatives ever undertaken. Fifteen years earlier at the United Nations (UN) headquarters in New York, world leaders convened to launch a set of Millennium Development Goals (MDGs) in order to rescue the "bottom billion." The term referred to an estimated one billion people who remained trapped in poverty, living on an income at or below the World Bank poverty level, which at that time was $1 per day.

Two of the eight MDGs targeted infectious and tropical diseases, recognizing that these conditions actually perpetuate poverty because of their ability to either devastate families or cause debilitating health effects. One goal focused on killer childhood infections preventable by vaccines, while the other aspired to combat "the big three infections": HIV/AIDS, tuberculosis, and malaria. Later, I worked with colleagues to raise awareness about a group of chronic parasitic and other related illnesses, branding them as neglected tropical diseases, or NTDs, and added these conditions to the big three [1]. A 15-year timetable was set to achieve the MDGs.

Global Infectious Diseases

Among the different categories of global infectious diseases, the neglected tropical diseases (NTDs) refer to a group of approximately 20 chronic and debilitating tropical infections, such as hookworm infection, schistosomiasis, leishmaniasis, and Chagas disease. I use the broader term "neglected diseases" to refer to a category that combines the NTDs together with HIV/AIDS, malaria, and tuberculosis, while "emerging diseases" mostly refer to newly arising viral illnesses such as Ebola, COVID-19, or Nipah virus infections. In some cases, certain neglected diseases such as malaria or leishmaniasis can also emerge in times of war or conflict or for other reasons that lead to the collapse of health systems. Yet another category is the major vaccine-preventable diseases of childhood, such as measles, polio, diphtheria, pertussis, and tetanus, to name a few. These disease categories represent the illnesses that are now returning after 2015, and these are the ones we need to target through vaccine diplomacy.

Progress through the MDGs: Disease Reductions, 2000–2015

To advance the Millennium Development Goals, key global leaders such as President George W. Bush (Bush 43) and Prime Minister Tony Blair mobilized large-scale funds and resources to combat the most pressing diseases affecting the poor. For the neglected diseases, ambitious programs of specific mass drug treatment interventions—including the US President's Emergency Plan for AIDS Relief; the Global Fund to Fight AIDS, Tuberculosis, and Malaria; and the US President's Malaria Initiative—started to produce important reductions in the number of people affected by these serious infections [2].

In the early 2000s, I lived and conducted research on NTDs in Washington, DC, where I was then chair of the Department of Microbiology and Tropical Medicine at George Washington University. However, I also used my time in the nation's capital to make frequent visits both to the executive office buildings of the White House and congressional office buildings in order to convince leaders about the opportunities to target NTDs with existing medicines. The Bush 43 administration and Congress were receptive and ultimately appropriated funds through the US Agency for International Development to support the distribution of packages of essential medicines for NTDs, mostly parasitic worm infections and trachoma. The medicines could be administered once annually for less than US$1 per dose. Over time, hundreds of millions of people began receiving essential medicines for NTDs, and in some countries, we even saw steep declines of selected NTDs, such as lymphatic filariasis (also known as elephantiasis), onchocerciasis (river blindness), and blinding trachoma, to a point where we could envision their elimination, referring to an absence of ongoing disease transmission.

The impact of these global health programs was especially apparent for sub-Saharan Africa, where disease reductions were so profound that there was evidence they caused a geographic shift in poverty-related diseases. On a global scale, I found that by 2015 most the world's poverty-related neglected diseases, including the NTDs and even AIDS, malaria, and tuberculosis were no longer mostly the diseases of Africa, but instead they were increasingly found in the pockets of extreme poverty that remained in mostly wealthy nations [2]. This represented a switch in the typical norms of global health, which typically juxtaposed developed versus developing countries. In its place arose a new paradigm linked to the fact that most economies were rising but leaving behind impoverished individuals affected by poverty-related diseases. I found that a decade and a half after the Millennium Development Goals were launched, the poor living in wealthier Group of 20 (G20) economies now accounted for most of these diseases.

Even more impressive were the reductions in vaccine-preventable childhood infectious diseases. In 2000, the Bill & Melinda Gates Foundation helped to fund and create an extraordinary partnership known initially as the Global Alliance of Vaccines and Immunizations, and since called Gavi, the Vaccine Alliance. The major remit of Gavi was to expand vaccine access for children living in the poorest regions of Africa, Asia, the Middle East, and Latin America, while simultaneously introducing new vaccines for rotavirus diarrhea and pneumococcal pneumonia and meningitis. These activities ignited a tremendous expansion in the number of children receiving essential vaccines.

The impact of vaccines on childhood diseases was even more dramatic than the mass treatment of neglected diseases. We recently learned about the public health gains achieved as a result of vaccinations through an evaluation conducted by a group of hundreds of investigators linked to the Global Burden of Disease Study (also supported by the Gates Foundation). Between 2000 and 2017, the numbers of children under the age of five dying from measles dropped from almost 500,000 deaths annually to under 100,000 deaths, for the first time. Many children still die of measles, but Gavi and its partners have achieved an 80% decrease in deaths since the launch of the Millennium Development Goals. Similar or dramatic percentage reductions also occurred in the deaths from other vaccine-preventable diseases, including diphtheria, pertussis (whooping cough), tetanus, and *Haemophilus influenzae* type b.

Particularly sharp declines in two vaccine-preventable diseases stand out. Now, for the first time, we could envision the potential elimination of ancient childhood scourges such as polio and measles. In the decade before the Millennium Development Goals, polio was endemic in more than 100 nations, but by 2019 it had decreased to a point where only 3 countries—Afghanistan, Nigeria, and Pakistan—were currently experiencing regular transmission of this disease. Equally impressive were the consequences of global vaccinations against measles. Measles is one of the most highly contagious diseases known. The measles virus has a reproductive number of 12–18. This means that an individual with measles on average will infect at least a dozen other people, typically infants less than one year of age who are not yet old enough to receive their vaccination. The high level and ease of transmissibility means that the control or elimination of measles in a community usually requires that we vaccinate at least 90 to 95% of that population. Essentially, almost everyone has to receive vaccinations. The fact that we achieved an 80% measles reduction means that Gavi and the disease-endemic nations were reaching most of the world's population at risk for this disease. As a result, there was great optimism that we might be on our way to stop transmission of measles, an almost unimaginable prospect just a few years before.

By the start of 2015, the global health community of scientists and public health experts were not exactly in celebratory mode, but there was an enormous sense of pride and optimism that we were on a path toward widespread elimination. Childhood vaccine-preventable diseases were going down, and we could envision future gains to the point where polio or measles might disappear in most of the developing world. Without a vaccine, the public health gains against AIDS, tuberculosis, malaria, and NTDs achieved through mass treatment were not as impressive, but they were still substantial. Global infectious diseases seemed on their way out, and policymakers began projecting steep declines in both neglected and vaccine-preventable diseases. Our global optimism about the disappearance of global infections was tempered only by projections that noncommunicable illnesses, such as cancer, cardiovascular disease, and diabetes were increasing. The Global Burden of Disease Study projected that the rise in these noncommunicable diseases was roughly commensurate with the decline of poverty-related infections. With infectious diseases trending downward, diabetes, cancer, and heart disease might become the dominant global health challenge.

Unraveling: An Overview

Unfortunately, beginning around 2015, we started to see unexpected and fundamental changes leading to a new order in which infectious and tropical diseases either emerged or returned. This book focuses on some of the major twenty-first-century forces responsible for this historic reversal, but here is a brief overview of some of the major determinants now driving up both vaccine-preventable diseases and NTDs.

Political Instability. One of the most potent and unanticipated drivers was political instability. Measles was declared eliminated from the Americas in 2016, but in Venezuela, the collapse of the economy interrupted and disabled its health system, allowing measles to come roaring back. However, measles was not the only infection to reemerge. Malaria also became widespread, as did a host of other NTDs transmitted by insects or snails. The so-called Northern Triangle area of El Salvador, Guatemala, and Honduras also suffered because of escalating drug wars and the resulting economic downturns that affected health systems. In the Old World, wars or Islamic State occupation in Syria, Iraq, and Yemen also promoted the return of vaccine-preventable diseases, including measles and polio, while the simultaneous collapse of insect vector control programs promoted an explosion in the number of cases of cutaneous leishmaniasis, a highly disfiguring disease that produces ulcers and permanent and socially stigmatizing scars. A deadly cholera outbreak, one of the largest ever recorded, swept across Yemen. Throughout war-torn Democratic Republic of Congo (DR Congo), Central African Republic, and South Sudan, measles also returned, as did another form of leishmaniasis known as kala-azar, which causes a leukemia-like illness that killed thousands. Ebola caused a new lethal epidemic in DR Congo in 2019, resulting in over 2,000 deaths, and even more fatalities occurred from measles and cholera. The bottom line was that new twenty-first-century wars, conflict, and political unrest were reversing global gains.

Internal Displacement and Human Migrations. Exacerbating war and political instability were the ensuing human migrations. As people fled conflict and political collapse, thousands of refugees poured into neighboring countries and regions to spread disease. Measles became widespread in Brazil, Colombia, and Ecuador, largely reversing the celebrated 2016 achievement of measles elimination in the Western Hemisphere. Populations also began fleeing the drug wars of the Northern Triangle, although this has not yet translated into measles epidemics. However, the disease did reemerge among displaced people in multiple African nations and in the Middle East. The World Health Organization (WHO) issued a global alert on measles and then followed it with a report revealing that approximately 10 million children in 16 countries were not receiving their routine childhood vaccines for measles, pertussis, and tetanus owing to conflict and human displacement [3]. Similarly, leishmaniasis traveled with the Syrian refugees spilling into Jordan, Lebanon, and Turkey, in some cases establishing a foothold in those countries.

Urbanization. Human migrations from conflict and other factors brought people in large numbers into cities in vast and unprecedented numbers. Thousands crowded into urban slums in Caracas (Venezuela), Aleppo (Syria), and Kinshasa (DR Congo). The urban slums of megacities became a dominant theme of a new world order. As populations outstripped infrastructures, diarrheal diseases, including cholera, emerged in the untreated sewage, while respiratory diseases, including measles and other vaccine-preventable infections, emerged in the crowded conditions. Then the coronavirus disease of 2019 (COVID-19) swept across densely populated urban regions of central China, and next Europe and the United States, ultimately causing a destructive pandemic that may trigger a new economic depression. COVID-19 now represents an imminent threat to vulnerable people living in the crowded urban slums of South Asia, the Middle East, Africa, and Latin America.

Anti-science and Nationalism. An equally worrisome social determinant was the new reality of anti-science. The anti-vaccine or anti-vax movement began to take off in the early 2000s, but by 2015, it had become an ugly monster. It emerged as a media empire, with by some accounts more than 400 misinformation websites actively promoted on social media and e-commerce sites. The anti-vax movement weaponized both Facebook and Amazon in their unique ways. Facebook became the major voice of the anti-vaccine movement, while Amazon turned into the greatest promoter of phony, misinformative books and documentaries. Then the movement acquired a political arm that created political action committees (PACs), each working to enact legislation that made it more and more difficult for children to receive access to vaccines. In 2015, a PAC in Texas arose from the Tea Party, a far-right-wing element of the Republican Party [4]. A similar anti-vaccine initiative steeped in the rhetoric of populism arose in Italy. Somehow the anti-vax movement became tied to a new nationalism arising in the United States and Europe. Nationalism itself became a social determinant of disease.

Later in 2017, the leaders of the anti-vax movement began engaging in predatory behaviors to target selected ethnic and religious groups. As vaccination coverage declined among both Somali immigrants in Minnesota and orthodox Jewish communities in New York as a result of specific targeting by the anti-vax movement, terrible measles outbreaks ensued in 2017 and 2019, respectively. Ultimately, measles epidemics became widespread across North America, while Europe suffered a record 80,000 cases in 2018 and 90,000 cases in the first-half of 2019. Despite the great gains from Gavi, measles reestablished a foothold in the United States and Europe. Epidemics also surfaced in Philippines, Samoa, Madagascar, and elsewhere in the developing world, to the point where the WHO declared "vaccine hesitancy" as one of the world's most pressing global health issues.

Climate Change. The new twenty-first-century determinants of disease also went beyond social ones. Climate change became a dominant force promoting disease. Mosquito-transmitted arbovirus illnesses such as Zika virus infection, chikungunya, and dengue spread across Central and South America and the Caribbean, before entering Texas and Florida in the United States. In southern Europe, West Nile virus infection and other arbovirus illnesses became common; malaria reappeared in Greece and Italy after it had been gone for decades; and schistosomiasis emerged on the island of Corsica. The Middle East experienced unprecedented high temperatures, which often exceeded 50°C, together with periods of severe and prolonged drought, forcing many to abandon their ancient agricultural lands.

However, it was difficult to attribute the appearance or reappearance of these tropical infections unambiguously to climate change. As noted above, in both the Western Hemisphere and southern Europe, human migrations were also widespread, linked to diaspora from Venezuela and the conflict zones of the Middle East and North Africa, respectively. Cities became vast, crowded, and susceptible to infectious disease transmission. If this trend continues, by 2050 the world will be constituted mostly of hot and steamy megacities, each with more than 10 million people. Complicating things further were the sharp economic downturns in many of these cities, especially in Venezuela, Brazil, the Middle East, and parts of southern Europe. COVID-19 furthered these economic declines. In other words, climate change went hand-in-hand with refugee movements, urbanization, and economic collapse. We had no real way to accurately attribute the risk to the individual social and physical determinants that are bringing back global tropical infectious diseases. However, one thing was clear: diseases that we thought we had vanquished through programs of the Millennium Development Goals were now returning.

Science Envoy

My term as US science envoy coincided with the rise in these geopolitical forces and climate change. I focused on evaluating the diseases arising from the conflict zones and then on designing new technologies to prevent these illnesses. As codirector of a nonprofit organization developing vaccines to combat NTDs (Texas Children's Hospital Center for Vaccine Development), and as someone with expertise in tropical infectious diseases (I am dean of the National School of Tropical Medicine at Baylor College of Medicine), I had a unique perspective on the diseases arising in this part of the world. In time, we redirected some of the activities of our laboratory toward making vaccines to combat some of the leading illnesses. They included vaccines for leishmaniasis, schistosomiasis, and the major coronavirus infections, including Middle East respiratory syndrome (MERS), a highly lethal disease. We were positioned to assist in building capacity for vaccine development and clinical testing across the Middle East. Vaccines are not the only tools needed to fight the emerging and neglected diseases arising out of the conflict zones, but they are perhaps the most efficient and effective at preventing disease. Yet the Middle East and North Africa are highly depleted in terms of vaccine development capacity. At the time I began as US science envoy, these areas possessed few to no vaccine development capabilities. Moreover, the major pharmaceutical vaccine manufacturers had little interest in developing vaccines to combat the neglected and emerging diseases of Syria, Iraq, and Yemen, and at best modest interest in vaccine capacity building. I therefore embarked on a journey in vaccine diplomacy in order to combat the infections arising in the post-2015 new world order, guided by the example of my role model in this endeavor, Dr. Albert Sabin.

2:
A Cold War Legacy

I never had the opportunity to meet Dr. Albert Sabin. He passed away in 1993 before I began my association with Sabin Vaccine Institute, a Washington, DC, nonprofit organization that advocates for vaccines and vaccine science. However, for more than 20 years I was connected with the institute. My association began when I was on the Yale faculty (the institute was started by H. R. Shepherd, a businessman based in New Canaan, Connecticut); continued during the 11 years when I was microbiology chair at George Washington University (the institute relocated with me in 2000); and then ended after I had relocated to Houston, Texas.

One of my favorite activities as president of the Sabin Vaccine Institute was visiting Dr. Sabin's widow, Heloisa. Heloisa lived off New Mexico Avenue in Washington, DC, not far from the campus of American University. She was born in Brazil and worked at *Jornal do Brasil*, the major newspaper in Rio de Janeiro. By the time Heloisa met Sabin at a reception for him in Brazil, both had been married previously. Shortly after their marriage in 1972, Heloisa moved with him to Israel when he served as president of the renowned Weizmann Institute, before moving to Washington, DC.

Heloisa's New Mexico Avenue apartment was like a mini-museum to vaccine diplomacy. It featured pictures of Sabin with President Clinton, Pope John Paul II, and Cuba's Fidel Castro, to name a few. She also had photos of Sabin with Soviet scientists, and on the tables and walls were plaques and remembrances from dozens of countries. Typically, after sitting in her apartment we would go downstairs to have lunch in a restaurant located nearby. We would pass the time talking about Sabin's life, his fierce determination to vaccinate the world's children against polio, and the many complexities of working with foreign governments to conduct vaccination campaigns. One story I remember vividly was her account of Sabin's visit to Brazil in 1980, when he had openly criticized federal and local health officials for their handling of a polio outbreak. Ultimately, his offer to help Brazil mount a national polio campaign was rebuffed, and Sabin returned disappointed to Washington. There were differing accounts of whether the Brazilian officials were too lax or if Sabin was too abrasive; possibly both were true [1]. Sabin was known for his directness, and his unrelenting demand for excellence often made people around him uncomfortable, but Heloisa both adored and revered him. She was petite and beyond charming, and from her pictures I could tell that back in the day, Heloisa and Albert were probably quite the glamorous couple. Heloisa would always refer to him as "my Albert." On a few occasions, we would visit his gravesite at Arlington National Cemetery, and she would always remind me that one day she would be buried alongside him. Heloisa passed away in 2016 in her late 90s, just before I left the Sabin Vaccine Institute. Currently, the Albert B. Sabin Archives are located at the University of Cincinnati, where he conducted much of his path-breaking work on the oral polio vaccine.

Sabin was a champion of vaccines, but not only because of his important and fundamental research to develop the polio and other vaccines. He was an unofficial polio ambassador, visiting dozens of countries and convincing government leaders at the highest levels about the importance of instituting polio vaccination campaigns. His stature as a vaccine scientist allowed him entry into Cuba during the 1960s and the USSR in the 1950s and 1960s. Those activities in Cuba and the USSR had special meaning for me. Through a program of back-channel diplomacy and scientific collaboration, Sabin worked with Soviet scientists to jointly develop an oral polio vaccine that employed Sabin's live virus polio strains, which he had first developed at Cincinnati Children's Hospital. Those virus strains were then produced at an industrial scale in the USSR and tested on millions of Soviet citizens, ultimately leading to the licensure of the vaccine in the early 1960s and the subsequent eradication of polio. These accomplishments are now the gold standard for how scientists of different ideologies can overcome diplomatic tensions or even overt conflict in order to advance science for humanitarian purposes.

Global Health Diplomacy

Each visit with Heloisa reinforced my conviction that vaccine diplomacy could one day hold a special place in modern society. In our post-2015 world, we need vaccine diplomacy more than ever. Global infectious diseases have taken an unexpected turn for the worse. Owing to breakdowns in health infrastructure from war and instability, together with other modern twenty-first-century forces, infectious diseases once thought to be on their way out, or even gone, are now back. The COVID-19 pandemic is testing international relations on an unprecedented level. Solving these and future infectious disease public health crises will require us to integrate the science of tackling global infections with these new social and physical determinants: poverty, war, political instability, human migrations, urbanization, and anti-science. In turn, navigating such troubled waters will require new approaches linking biomedical and social sciences, including political science and foreign policy.

In my two years in the Obama administration as US science envoy, I came to realize that understanding the biomedical science, the vaccinology, was essential but not always sufficient to solve issues related to building vaccine infrastructures across nations. This was especially true in a complicated space like the Middle East, where deep-seated tribal and Sunni-Shia rivalries continuously threw up roadblocks—often in interesting and unexpected ways. It became apparent that building vaccines, expanding vaccine coverage, and tackling NTDs requires integrating new types of knowledge, including skills related to diplomacy. In some ways, this might bear some resemblance to what Sabin achieved in Cuba and the USSR (okay, maybe not Brazil!) in the 1960s, but widening the tent to include both scientists and nonscientists. To achieve this, I suggested a new framework of vaccine diplomacy that connects political science, philosophy, and foreign policy to the most powerful life science technology ever invented—vaccines.

Before describing and defining vaccine diplomacy, I think it is helpful to first provide a broader understanding of how global health in general is linked to international relations and solving disease problems on a large scale [2]. Some might say it began as an early version of quarantine during the 1300s, when laws were implemented to prevent plague originating in Asia Minor from entering Dubrovnik on Croatia's Adriatic coast—or much later, starting in the 1850s, when international sanitary conferences were held in Europe to prevent cholera, plague, and other pandemic infectious disease threats from spreading [2]. Then, in the early twentieth century, the Office International d'Hygiène Publique was created in Paris, as well as a health organization linked to the League of Nations [3]. In parallel, the nations in the Western Hemisphere also established a Pan American Sanitary Bureau, later named the Pan American Health Organization, which became the regional office of WHO in the Americas. The actual World Health Organization itself was established in the aftermath of World War II, following the formation of the UN. The WHO's constitution was enacted on April 7, 1948, now designated as World Health Day. Almost twenty years later, the WHO embarked on the eradication of smallpox through a global vaccination campaign.

Global health diplomacy rapidly accelerated after promulgation of the UN's Millennium Development Goals, first in 2005, with a revised set of International Health Regulations (IHR), and then in 2007, after the ministers of health of seven nations connected global health to foreign policy through an Oslo Ministerial Declaration [2]. IHR, also known as IHR (2005), is an agreement between all WHO member states focused on global health security, especially for the detection and assessment of major public health events and for strengthening disease control efforts at national entry points, such as seaports and airports. A key driver of the IHR (2005) was the 2003 pandemic of severe acute respiratory syndrome (SARS) that resulted in more than 8,000 cases, with roughly 10% mortality [4]. The SARS pandemic also severely affected the economies of Hong Kong and Toronto, Canada, and were a wake-up call for the disruptive power of lethal epidemics. These initiatives were later strengthened in 2019 following the Ebola epidemic in DR Congo and ultimately were called on to respond to COVID-19 the following year. In this context, my former Yale colleague, Ilona Kickbusch, defines global health diplomacy as a system of global governance in health, while Rebecca Katz, a colleague and former student now at Georgetown University, provided an operational definition. She refers to it as a framework to include treaties between nations—such as IHR, or recognized international partnerships with UN international organizations, Gavi, or global partnerships involving the Gates Foundation or other non-state actors [2].

Vaccine Diplomacy

Throughout modern history, vaccines have surpassed all other biotechnologies in terms of their impact on global public health. Because of vaccines, smallpox was eradicated, and polio has been driven to near global elimination, while measles deaths have declined more than 90%, and *Haemophilus influenzae* type b meningitis is now a disease of the past in the United States and elsewhere.

I define one part of vaccine diplomacy as a subset or specific aspect of global health diplomacy in which large-scale vaccine delivery is employed as a humanitarian intervention, often led by one or more of the UN agencies, most notably Gavi, UNICEF, and WHO, or potentially a nongovernmental development organization [2]. Examples might include emergency cholera or Ebola vaccinations during outbreaks in Africa, measles vaccination campaigns linked to the Venezuelan diaspora in Brazil or Colombia, or polio eradication campaigns in the conflict areas of Afghanistan, Pakistan, or the Middle East. Other aspects of vaccine diplomacy relate to vaccine access during pandemics, such as efforts to ensure equitable delivery of a vaccine to combat influenza, especially during an epidemic or even a pandemic situation.

Another critical element of vaccine diplomacy includes the development or refinement of new vaccines achieved jointly between scientists of at least two nations. Rather than a UN agency or nongovernmental development organization, the actual scientists lead both the vaccine science and diplomacy [2]. It is especially relevant that scientists from nations in opposition or even outright conflict can work in research organizations, or that they are able to work together and engage in collaborations under conditions of political instability or stress. Under this definition, vaccine diplomacy reached its full expression during a 20-year period of the Cold War between the United States and Soviet Union that began around the time of the Sputnik satellite launch and mostly ended in 1977 with the eradication of smallpox [5]. In my role as US science envoy, I worked to resurrect this vaccine science diplomacy while collaborating with scientists from Muslim-majority countries of the Middle East and North Africa [6].

Do vaccines really deserve their own designation for a special type of diplomacy? Yes, I believe so, especially when we consider that between the past century and this one vaccines have saved hundreds of millions of lives [2]. In this sense, the technology of vaccines and their widespread delivery represent our most potent counterforce to war and political instability in modern times. Vaccines represent not only life-saving technologies and unparalleled instruments for reducing human suffering, but they also serve as potent vehicles for promoting international peace and prosperity. They are humankind's single greatest invention.

A Brief History of Vaccine Diplomacy

The history of vaccine diplomacy traces an interesting narrative parallel to the history of the vaccines themselves. It started with the British physician Edward Jenner, who in the late 1700s developed the first and original smallpox vaccine. Indeed the word *vacca*, Latin for "cow," refers to the fact that the attenuated virus used in the vaccine derived from cows infected with the cowpox virus. However, a more recent analysis questions the true origins of the virus that Jenner actually used, which might have been horse pox, or even another virus entirely, designated simply as "vaccinia" [7]. Jenner's smallpox vaccine had an immediate impact on public health in England, and it was transported across the Atlantic Ocean to America, where Thomas Jefferson himself conducted vaccine trials in and around Virginia. When he commissioned Meriwether Lewis and William Clark for their expedition a year after the Louisiana Purchase in 1803, Jefferson either encouraged or arranged for them to carry the vaccine into the frontier [2]. Back then, smallpox was devastating Native American populations in the Northern Plains, so the vaccine was potentially a gesture of peace or goodwill. Unfortunately, some historians report that the vaccine preparation degraded to a point where it was never actually used.

In Europe, both England and France celebrated and honored Jenner's achievements despite increasing hostilities between the two nations. In the period following the French Revolution and after Napoleon became military dictator of France in 1799, Britain had become increasingly concerned about his armies expanding across Europe and his efforts to stop European nations from trading with England. Finally, in 1803, Britain declared war on France, beginning with a naval blockade of the country. Historic battles at Austerlitz and Trafalgar ensued. However, Jenner's reputation and veneration as the first vaccine scientist had grown to such a point that he was asked to write letters (and possibly engage in other activities) mediating the releases or exchanges of prisoners [2]. For example, in a letter to the French National Institute of Health, he asserted that "the sciences are never at war." In turn, Napoleon (or some say Empress Josephine) declared, "Jenner—we can't refuse that man anything" [2]. Ultimately, the Napoleonic wars ended with Napoleon's defeat at Waterloo, the last time France and England went to war.

These vignettes highlight a future paradigm that subsequently held for the next 200 years—namely, (1) the immediate recognition of the impact of a vaccine as a highly valued technology, and (2) the enormous scientific and professional stature of vaccinologists—vaccine scientists and vaccine developers. That is, until the modern-day anti-vaccine movement began to target us beginning in the early 2000s. An elusive third element, although one not as straightforward and tangible, also attaches to vaccines: the potential for vaccines to both prevent diseases arising out of conflict or twenty-first-century forces, and in some cases, to directly address the actual social determinants. For example, Jenner's vaccine was itself employed as an instrument of peace during the Napoleonic wars, creating a novel thread through modern history. When another renowned Frenchman, Louis Pasteur, developed the next few vaccines in the mid-1800s, he also used his stature to launch a network of Pasteur Institutes across the Francophone world, including North Africa and Southeast Asia, which initially focused on reproducing Pasteur's method to prepare and deliver the first rabies vaccine. Echoing Jenner's comments, Pasteur in an 1888 speech at the founding of the Institut Pasteur in Paris remarked that "science has no country, because knowledge belongs to humanity and is the torch which illuminates the world" [2].

The Cold War was a 45-year period of political hostilities between the United States and the Union of Soviet Socialist Republics that began after World War II and divided much of the globe into two major spheres of influence. Ironically, it became the signature period that generated the fullest expression of vaccine diplomacy. Two enemies put aside their animosities in order to collaborate on the development and testing of the oral polio vaccine, which is now leading to its global elimination or eradication. This is an extraordinary story that few people outside the vaccine world know about. The 1957 launch of the Sputnik satellite was a key moment in American history, when the nation feared falling behind the Soviets in mastery of both space and missile technology. It became a dark chapter in US history when—following on the heels of the "red scare" that resulted from the Soviet annexation of eastern Europe, the Berlin blockade, and our proxy war with China in Korea—we became vigilant, even hyper-vigilant, for any signs of Communist presence on our soil.

One might argue that this was not an ideal time to begin a US-Soviet scientific collaboration on vaccines, but that is more or less what occurred. It turned out that the fear of polio exceeded the threat of Communism. The 1952 polio epidemic in America was the worst on record. It killed more than 3,000 people and caused partial or complete paralytic disease in more than 20,000. Increasingly during the 1950s, school-age children and adolescents were polio victims. Parents in cities across America lived in terror of summer polio epidemics.

Polio also raged in the USSR. Between 1954 and 1959, polio was present in all of the Soviet republics and increasing in incidence yearly, with the highest rates in the Baltics [8]. Polio outbreaks also occurred in Moscow and Minsk [9]. In response, Soviet scientists in 1955 established a Poliomyelitis Research Institute in Moscow and appointed Dr. Mikhail Chumakov to head experimental vaccine development. Another key individual was Dr. Anatoly Smorodintsev, the head of the virology department of the Institute of Experimental Medicine of the USSR Academy of Medical Science [8]. With the agreement of both governments, Chumakov and Smorodintsev traveled to the United States in 1956 to visit with Albert Sabin, who had developed a polio vaccine containing three different live, attenuated poliovirus strains administered by mouth.

Sabin was eager to cooperate with the Soviet scientists because by this time the injectable polio vaccine invented by Dr. Jonas Salk, composed of three virus strains that have been inactivated or killed with formalin, was already licensed and widely used in the United States. So not only was there little appetite to replace it with the Sabin vaccine, but there were insufficient numbers of unvaccinated American children available for testing with the oral version [9]. While Dr. Sabin was able to immunize his own family and small numbers of incarcerated young adults at a federal prison located not too far from his Cincinnati Children's Hospital laboratory, the number of vaccinated volunteers was far too small for his oral vaccine to achieve product licensure in the United States [9].

As an aside, I will mention that I was privileged to meet with Jonas Salk in 1995. Our meeting was held in his office at the Salk Institute, considered by many to be one of the most visually striking research institutes ever built. It was designed by the famed architect Louis Kahn and overlooks a beach on the coast of La Jolla, California, just north of San Diego. At sunset, the

The Salk Institute. Photo by Wikimedia Commons user TheNose, https://creativecommons.org/licenses/by-sa/2.0/deed.en.

light shines through a gap between its two major buildings, providing an unforgettable effect. At that time, I was an assistant professor at Yale, just a few years into beginning my own vaccine laboratory. Dr. Salk was one of the most gracious and welcoming senior scientists of stature I had ever met. He even agreed to help me further the development of our hookworm vaccine. We spent more than an hour together, during which he proudly showed me the paintings he displayed in his office by his wife, Françoise Gilot (and former partner of Pablo Picasso). I remember the exhilaration after leaving our meeting, believing that Dr. Salk could become an important mentor. I was devastated when just a month later my wife, Ann, phoned me at a meeting in the United Kingdom to tell me that he had passed.

To return to the story: Because the Salk vaccine required an injection, there was still a global need for a different type of polio vaccine that could be administered without the use of a needle and therefore trained medical personnel. This was especially important in developing countries of Africa, Asia, and Latin America, where qualified staff were lacking and health systems were depleted. For poor nations in Africa, Asia, and the Americas, the Sabin vaccine checked a number of boxes. It contained live polioviruses that were weakened or "attenuated" to a point where they can no longer cause disease. The advantage is that the Sabin poliovirus vaccine strains can be given orally because they stimulate a child's immune response by replicating in the gastrointestinal tract. If it worked, a large group of children in a village or town could be lined up and given the Sabin vaccine via liquid drops, or even drops placed on a sugar cube.

Following the 1956 visit by Russian scientists to the United States, our State Department allowed Sabin to make a reciprocal visit in the summer of that year [9], launching an extraordinary international collaboration in which Sabin's live polio strains were scaled up for production in the USSR and first tested in Soviet children. Sabin provided sufficient amounts of the vaccine to begin immunizing children in both the USSR and Czechoslovakia, as well as seed lots so that the Soviets, under the direction of Dr. Chumakov, could scale up the virus themselves. According to William Swanson, a freelance journalist based in Minneapolis who wrote about this period for *Scientific American*, Chumakov had to use his Politburo connections to go over the head of the Russian minister of health, who would not authorize clinical testing of the Sabin vaccine [9]. Chumakov was a courageous man, who always put science and the health of the USSR's children above politics. His son Dr. Konstantin Chumakov, himself an important vaccine scientist at the US Food and Drug Administration, is a colleague who shared with me fond remembrances of his father, who was a great defender of science during a very difficult period both before and immediately after the death of Stalin. Dr. Mikhail Chumakov died just a few years after Sabin passed away in 1993.

Ultimately, toward the end of 1959, the Russians had successfully prepared 10 million doses of the vaccine derived from Sabin's live polio strains. The Soviets vaccinated millions of children. In 1991, Dorothy Horstmann, one of the founding professors of virology at Yale University (and a former mentor who recruited me to Yale after my residency at the Children's Service of the Massachusetts General Hospital) wrote about her experience of providing an independent assessment of the subsequent polio clinical trials that began in 1959. In anticipation of those trials, the WHO asked her to assess in detail over a six-week period the quality control of the polio laboratories and whether the Soviets implemented adequate steps to ensure the safety of the vaccine [8].

She also reported on Chumakov's travels across the USSR to meet with groups of physicians in order to organize the vaccination campaign. He went on local TV and radio to ask for community cooperation and worked with local newspapers to explain the importance of the vaccination drive [8]. Detailed surveillance surveys followed, which included home visits by healthcare providers to assess whether there were any harmful effects of the vaccine. In the case of a campaign launched in Tashkent, Uzbekistan, during an actual polio epidemic, epidemiologists went to individual homes to determine the public health impact of the vaccine. An estimated 10–15 million Russian children received the oral polio vaccine developed jointly by Sabin on the US side and Chumakov and colleagues on the Soviet side. Ultimately, almost all Soviet citizens under the age of 20, approximately 100 million individuals, received the vaccine. According to Dr. Horstmann, "Its positive assessment contributed to a rebirth of interest in the oral vaccine and paved the way for large field trials in the United States, leading to licensure of oral vaccine in 1961–62" [8].

As a result of global access to the oral polio vaccine, by 2019 the disease has been eliminated from all but three countries—Nigeria, Afghanistan, and Pakistan—where local hostilities and conflict have interfered with efforts by UN agencies and community health workers to reach all of the areas that require vaccine access.

Downstream, the Sabin vaccine will likely be gradually replaced by the Salk vaccine. While it is clearly beneficial to administer an oral vaccine, a distinct disadvantage of the Sabin vaccine is that it is composed of live virus strains, which can undergo mutation. This means that vaccinated children can—although rarely—shed a mutated version of the virus into the community that produces complications similar to the wild-type poliovirus, including paralysis. Therefore, to truly eliminate global polio, it is believed that follow-up vaccinations with the Salk killed vaccine may be required, and the Salk vaccine is gradually being adopted in most countries. I find it ironic that both the Sabin and Salk vaccines were ultimately required to eradicate global polio. The irony stems from the fact that they were bitter rivals. In my Baylor College of Medicine office, I have a reproduction of a

Drs. Albert Sabin (*left*) and Jonas Salk (*right*).
Courtesy the Hauck Center for the Albert B. Sabin Archives, Henry R. Winkler Center for the History of the Health Professions, University of Cincinnati Libraries.

photo of Drs. Sabin and Salk seated next to each other at a conference, but what makes it special is the onlooker seated behind them with an expression of astonishment.

Health as a Bridge to Peace

Joint international development and testing of an oral polio vaccine proved to be a powerful force in overcoming Cold War ideologies. It also became one of the most important and successful biotechnologies ever invented—one that is leading to global polio elimination. Now the WHO has taken the humanitarian dimension of the oral polio vaccine a step higher through its Health as a Bridge for Peace program and its Humanitarian

Cease-Fires Project [10]. The project brokers cease-fires in war-torn areas of Afghanistan, Iraq, South Sudan, and elsewhere in order to vaccinate children against polio. A product of the Cold War, to this day the polio vaccine remains a potent weapon for waging peace and eliminating disease.

3: Vaccine Science Envoy

The legacy of US-Soviet vaccine science diplomacy carries on today through global polio eradication efforts and with the establishment in 1988 of an international Global Polio Eradication Initiative through a resolution passed that year by the World Health Assembly [1]. The annual World Health Assembly has a critical function in global health. It is where the world's health ministers meet at WHO headquarters in Geneva, Switzerland, each May to adopt new resolutions for addressing global health threats. For example, IHR (2005) was created at the World Health Assembly in the aftermath of the SARS epidemic. Ideally, such resolutions are followed by global efforts to mobilize resources—funds and human capital—to translate policy into action and to develop and implement timely interventions.

Over the years, I have been to several World Health Assembly events. They rank among the most important venues for shaping policies, while also affording opportunities to meet with high-level policymakers. Participation at World Health Assemblies has allowed me to press for interventions against neglected tropical diseases, to elevate the profile of global health research and development, and to combat global vaccine hesitancy. Most recently, in 2019 I spoke on a panel together with US Department of Health and Human Services Secretary Alex Azar to explain the growing threat of the anti-vaccine movement in America and my concerns about its potential to globalize. There is often a lot of energy and excitement at these venues, in part because key global health leaders, including those heading major UN agencies and nongovernmental development organizations, attend and actively participate. The press is often present as well. The assembly also helps to set future agendas and introduce key important new issues to the international stage.

When the World Health Assembly convened in 1988, polio was occurring in more than 100 nations. The response to deliver and expand global coverage of the oral polio vaccine and ultimately halt paralytic polio has been impressive. Over the past 30 years, the Global Polio Eradication Initiative has immunized more than 2.5 billion children, to the point where polio transmission continues in only three nations—Afghanistan, Nigeria, and Pakistan [1]. Moreover, the global incidence of polio has decreased by 99% since 1988, with two of the three polio types (types 2 and 3) now eradicated [1]. So far, the international investment in the Global Polio Eradication Initiative exceeds $10 billion. It includes contributions from donor countries but also from the Gates Foundation, along with extraordinary commitments from Rotary International, which comprises more than 35,000 service clubs and over one million members [2]. I recently became a Rotarian, and I have been fortunate to meet many others. Often, Rotarians are individuals with no medical background, but they are quick learners and great advocates who are eager to explain to lay people in their communities about the opportunity to eradicate polio.

The endgame in terms of going the "last mile" for polio eradication—preventing the final 0.1% of cases—will not be easy, because many children infected with the poliovirus are without symptoms. This reality regarding the natural history of polio generally means that finding a single case of a child paralyzed with polio often represents the tip of the iceberg, because on average there might be at least 100 other children who are infected with polio but who are asymptomatic. This makes ring vaccination, an approach developed during the eradication of smallpox in which contacts (or the contacts of the contacts) of a clinical case of smallpox are vaccinated, not generally possible. Instead, it becomes necessary to vaccinate entire villages or towns.

Another problem is that the live virus strains in the polio vaccine can mutate to cause vaccine-derived poliovirus (VDPV). This happens because a child vaccinated with the live polio vaccine can excrete or shed the live virus vaccine in feces. In areas of poor sanitation and where a low percentage of children are immunized, the vaccine virus can survive in the environment and undergo genetic mutations [3]. On rare occasions, those mutations can produce a live vaccine strain that has recovered its ability to cause paralytic polio. This is an extremely uncommon occurrence, but given that over the past two decades at least 10 billion children have received the oral polio vaccine, there have been hundreds of cases of VDPV [3]. Now that we are achieving almost complete global coverage with the oral polio vaccine, the hope and expectation is that VDPV cases, like wild-type polio virus, are on their way out. And as more and more countries adopt the Salk polio vaccine, which contains only killed poliovirus, we may one day finally eradicate this disease.

V.2.0: Global Smallpox Eradication Initiative

The Cold War collaboration between the American and Soviet scientists went beyond polio vaccinations. During the 1950s, the virologist and Soviet deputy health minister Dr. Viktor Zhdanov became concerned about smallpox importation into the USSR republics from Asia, especially its central Asian republics [4]. In 1958, he proposed to the World Health Assembly, then meeting in Minneapolis, Minnesota, that the WHO should consider launching an initiative to eradicate smallpox and pledged the delivery of a Soviet-made freeze-dried smallpox vaccine [4, 5]. Like the Sabin vaccine, the smallpox vaccine was also a live virus vaccine, but one that was administered by injection rather than orally. The advantage of freeze-drying the smallpox vaccine was that it enabled this live virus to withstand extremes of temperatures so it could be delivered to hot areas of the planet, making it possible to vaccinate populations living in tropical developing nations, such as India or the Sahel in sub-Saharan Africa.

I see Dr. Zhdanov as an unsung hero in the history of global health. He was trained as a military physician but became interested in epidemiology, ultimately becoming chief of epidemiology at an institute in his native Ukraine. He then became deputy minister, and it was in that role that he pressed the WHO to undertake smallpox eradication. In reference to Zhdanov, a 2015 article declared that "the best person who ever lived is an unknown Ukrainian man" [5].

Zhdanov had made a compelling proposal, but initially it met with a tepid response from the global health community. There were several reasons for resistance, including widespread skepticism about the feasibility of universal vaccinations [4]. Then there were also lots of complicated Cold War geopolitics. Some of the world's highly populated Communist nations, including China, had not joined the UN (or the WHO), and there were concerns about Soviet leadership and the potential for Communist influence on global health politics. More important, many African nations at that time were still colonies of European countries and therefore excluded from adequate WHO representation [4].

With only modest contributions coming from the WHO to support smallpox eradication, by the early 1960s the USSR was mostly going it alone in terms of scaling up production of the freeze-dried smallpox vaccine and then exporting it to vulnerable nations. Once again, the isolation of the USSR partly reflected Cold War politics. According to Dr. Donald A. Henderson (everyone referred to him as "D.A."), who went on to become the leader of smallpox eradication, the United States at that time was more committed to malaria eradication based on DDT insecticides and the antimalarial drug chloroquine [4]. Even though those global malaria eradication efforts ended in failure, owing to emerging insecticide and antimalarial drug resistance, they still consumed an enormous percentage of the WHO's budget and staff [4].

Despite the resistance of the WHO and the Americans, the Russians persisted and donated hundreds of millions of doses of smallpox vaccine to India and elsewhere [4]. Finally, in 1966, the World Health Assembly endorsed a global smallpox eradication program. It was proposed that the program would be headquartered at WHO and funded through international support. I do not have any insights on the negotiations that followed, but the Soviets, who led efforts to convince the WHO to focus on smallpox, unexpectedly gave the green light to allow D. A. Henderson, an American, to serve as director of smallpox eradication on behalf of the WHO. The USSR committed to donating 25 million doses of freeze-dried, Soviet-made vaccine annually. In the meantime, D.A. was invited to fly to the USSR in order to interview Russian physician-epidemiologists to work with him on the WHO smallpox eradication campaign [4]. A new chapter in US-USSR vaccine diplomacy was unfolding.

By 1973, the Russian donation of freeze-dried smallpox vaccine had reached the one billion doses mark [6]. According to Dr. William Foege, another giant of the smallpox eradication campaign who went on to become the director of the Centers for Disease Control from 1977 to 1983, the technology for making smallpox vaccine was also transferred to four facilities in

India for manufacturing their own indigenous freeze-dried vaccine [6]. By 1977, the last naturally occurring case of smallpox was found in Somalia, and after careful and exhaustive global surveillance activities, the world was declared smallpox-free in 1980 [7].

I find this to be a really extraordinary story of how the US, USSR, and India were able to work together in the 1960s and 1970s to solve a pressing pandemic threat like smallpox. Two decades after this achievement, I had the opportunity to sit down with D. A. Henderson in my microbiology chair's office at George Washington University. He was gracious enough to come and visit me, and it was a real honor to show him our laboratories, where at the time we were developing new vaccines for schistosomiasis and human hookworm infection. Following his time at WHO, he became the dean of the Johns Hopkins School of Public Health and then worked in Washington, DC, at the US Department of Health and Human Services to build biodefense initiatives together with my former mentor Dr. Philip K. Russell, ultimately leading to the formation of the Biomedical Advanced Research and Development Authority. On that day, D.A. and I discussed the power of vaccines, and it reinforced my sense that working decades to develop a new vaccine, especially for diseases of the poor, was a deeply meaningful path. I remember D.A. as one of the humblest, most self-effacing individuals I had met. During the entire time I spent with him, I had to keep reminding myself that I was meeting the man who led the initiative that eradicated the worst disease on our planet. During my term with the Sabin Vaccine Institute, I also had the honor to work closely with Dr. Ciro de Quadros, who reported to D.A. and led many smallpox vaccination initiatives globally. Eventually, Ciro joined the Sabin Vaccine Institute and became an internationally recognized expert on delivering vaccines. The Pan American Health Organization named Ciro a Public Health Hero of the Americas shortly before his death from cancer in

2014. He also became a great friend and colleague when I was president of the Sabin institute. I was at PAHO when he received his recognition to a thunderous standing ovation, something I will never forget. I was profoundly saddened when he died a few weeks later.

Smallpox was an illness that killed more people than all wars in the twentieth century, and D.A., William Foege, and Ciro de Quadros engineered one of humankind's greatest victories. It also occurred to me that the eradication of smallpox happened because of the availability of a freeze-dried vaccine first mass-produced in the USSR and delivered by a WHO program led by an American. In this sense, vaccine diplomacy produced two victories during and after the Cold War—global smallpox eradication and the near elimination of polo.

A New Beginning

While president of the Sabin Vaccine Institute during the 2000s, I became enamored with the idea that vaccine diplomacy could be more than just a historical curiosity or interesting Cold War tale, but that instead it was a concept to be resurrected.

The truth is that Russian-US vaccine diplomacy did not totally evaporate following the end of the Cold War and the break-up of the Soviet Union. There were productive bilateral initiatives focused on HIV/AIDS prevention, as well as other sexually transmitted diseases and tuberculosis, and in 2009, the United States and Russia created a Bilateral Presidential Commission devoted to both polio and malaria control, as well as health issues related to the consumption of tobacco and alcohol [8]. However, as I noted back in 2017, "these important efforts still fall short of the compelling stories offered by the joint vaccine science diplomacy that led to the oral polio vaccine now leading to global eradication efforts" [8].

Vaccine diplomacy literally received "a new beginning" on June 4, 2009, when Barack Obama in his first year as US president delivered a speech with that title at Cairo University in Egypt. The White House intended the speech to signal a new outreach to the Muslim world, under the premise that US-Arab relations had been severely damaged in recent history. The Obama White House considered Cairo to represent an important historic and political center of the Arab world, with Cairo University functioning as one of its key intellectual centers and where Taha Hussein, one of the greatest twentieth-century Islamic and Arabic scholars and intellectuals, had served as a professor and dean [9]. I have previously noted that Hussein also suffered from the neglected tropical disease blinding trachoma and lost his sight owing to a combination of the disease and failed eye surgery [9].

Barack Obama opened his speech by stating, "I've come here to Cairo to seek a new beginning between the United States and Muslims around the world, one based on mutual interest and mutual respect, and one based upon the truth that America and Islam are not exclusive and need not be in competition. Instead, they overlap, and share common principles—principles of justice and progress; tolerance and the dignity of all human beings" [10]. During the speech, the president emphasized the important intellectual and scholarly contributions of Islamic scholars in the areas of mathematics, navigation, architecture, poetry, music, writing and printing, and medicine. He spoke about how "bridges between peoples lead to action—whether it is combating malaria in Africa, or providing relief after a natural disaster" and then proposed an extraordinary new initiative in international science cooperation:

> On science and technology, we will launch a new fund to support technological development in Muslim-majority countries, and to help transfer ideas to the marketplace so they can create more jobs. We'll open centers of scien-

tific excellence in Africa, the Middle East and Southeast Asia, and appoint new science envoys to collaborate on programs that develop new sources of energy, create green jobs, digitize records, clean water, grow new crops. Today I'm announcing a new global effort with the Organization of the Islamic Conference to eradicate polio. And we will also expand partnerships with Muslim communities to promote child and maternal health. [10]

In the following days, I remember vividly reading the text of the speech. Even then, I realized it might one day have an important influence on my life. From my perspective, President Obama was openly speaking about vaccine diplomacy with the Muslim world, in ways that bore resemblance to Cold War vaccine diplomacy. With the end of the Cold War, I felt we needed a new frontier in vaccine diplomacy, and putting aside our ideological differences with Muslim majority nations to collaborate on vaccine science could become our twenty-first-century equivalent.

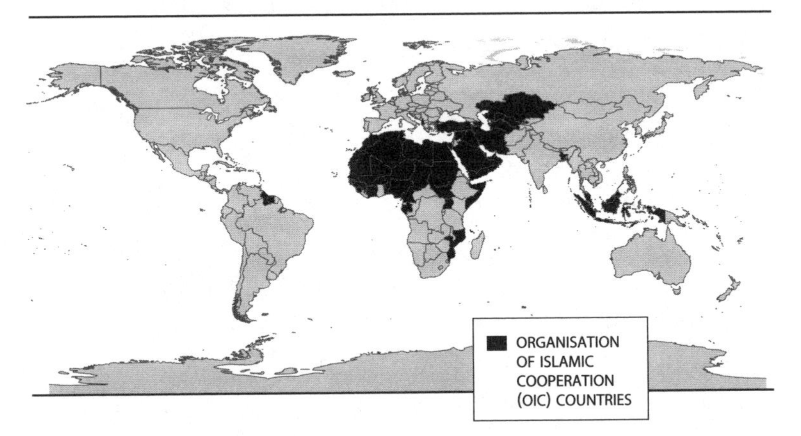

Organisation of Islamic Cooperation member states.
Courtesy of the Organisation of Islamic Cooperation. By Wikimedia Creative Commons user Mohsin, https://creativecommons.org/licenses/by/3.0/deed.en.

The speech President Obama delivered that summer inspired me to examine more closely the disproportionate suffering in the Muslim world, and in 2009 I observed how many neglected diseases occurred overwhelmingly in the Muslim-majority nations [11]. I found that NTDs such as schistosomiasis, hookworm infection, and leishmaniasis disproportionately affect the member states of the Organisation of Islamic Cooperation (formerly known as the Organization of the Islamic Conference), and it was not lost on me that my research group was working to develop new vaccines for these illnesses in our product development partnership laboratories. Because the long-term and debilitating health effects of NTDs trapped populations in poverty, the Muslim-majority nations urgently needed these "antipoverty" vaccines.

US Science Envoy Program

Obama's Cairo speech was followed up soon afterward with important actions. Just five months later, Hillary Clinton, then the US secretary of state, announced the establishment of a US Science Envoy Program during a speech in Marrakesh, Morocco, where a summit known as Forum of the Future was being held [12]. This became very much a bipartisan initiative, as it was a Republican, Senator Richard Lugar, who introduced the legislation in Congress for creating the program, before it was passed by the Senate Foreign Relations Committee [12]. A key component of both the legislation and announcement was the recognition that the United States has produced more Nobel laureates in science than any other nation and that our research institutes and universities are universally admired, often serving as the template or model for the creation of similar institutions globally. The creation of the US Science Envoy Program was the embodiment of something I had noticed at my 2015 dewaniya and indeed during my decades of international travel: America is loved because of its extraordinary scientific and intellectual horsepower, together with its willingness to train new generations of scientists.

Three outstanding scientists were chosen as the inaugural US science envoys: Ahmed Zewail, an Egyptian-born professor from Caltech who won the 1999 Nobel Prize in Chemistry; Elias Zerhouni, the former dean of Johns Hopkins University School of Medicine and director of the National Institutes of Health, who was born in Algeria; and Bruce Alberts, the past president of the National Academy of Sciences, editor in chief of *Science* magazine, and biochemistry professor and chair at the University of California, San Francisco [12, 13]. Working jointly with the US State Department, the White House Office of Science, Technology, and Policy, and US embassies abroad, the major tasks of the science envoys center on "sustained international cooperation" in the sciences. This includes advocating for transparency among scientific institutions in Muslim-majority nations, while promoting public engagement in science and science education, and advising US embassies in the host countries on opportunities for scientific exchange [13]. It also allows for the establishment of new science and technology projects. Still another important component is the outreach to promote the role of women in science abroad. Over time, there was a strong representation of women scientists among the envoys. The program officially began in 2010, and since then 18 envoys have visited more than 40 countries.

Vaccine Diplomacy in the Middle East and North Africa

In the fall of 2014, I was thrilled to receive a call from the US State Department inviting me serve a term as US science envoy. During the call, the State Department further indicated a specific interest in my focusing on countries in the Middle East and North Africa, both because of the mandates laid out in the 2009 speech in Cairo and the particular urgency arising out of complications from the growing threat of the Islamic State in Syria and Iraq.

However, within that space there were multiple opportunities to initiate vaccine diplomacy efforts. At the time, I was especially interested in promoting vaccine diplomacy with Iran. As I explained to the State Department and White House, initiating discussions on the vaccine diplomacy front with Iran bore some resemblance to the situation in the USSR that Albert Sabin faced in the late 1950s. First, there was an urgent need to develop new vaccines for the region, especially for diseases emerging out of the conflict zones, including vaccines we were developing for leishmaniasis and schistosomiasis. Second, Iran was one of the only nations in the region that actually had capabilities for producing vaccines through two major institutions—its Institut Pasteur based in Tehran and the Razi Vaccine and Serum Research Institute. An opportunity to work with already experienced Iranian vaccine scientists meant we could hit the ground running on one or more joint projects. Because Congress had not appropriated substantial funds for the US Science Envoy Program, it meant I needed to rely on the host country to provide a significant amount of in-kind support, and potentially funds could be mobilized by Iran's vaccine institutions. Finally, there was the exciting challenge of developing partnerships with a country where there were only modest levels (at best) of diplomatic contact, and where relations were strained. It was a situation that resembled how our country was interacting with the USSR during the 1950s and 1960s.

During our initial meeting in Washington, DC, we focused on selecting target countries for my role as science envoy. I spelled out the potential benefits and opportunities for a vaccine diplomacy project with Iran. However, while making my case, I immediately noticed that my audience's body language was not good—lots of folded arms—and I don't believe any questions were asked. It became apparent that I would not be going to Iran. Although I was a bit disappointed, especially because my State Department colleagues did not offer any particular reason at the time, we subsequently learned that the Obama administration had begun sensitive discussions with the Iranians regarding nuclear disarmament. I guessed that there was no way they were going to muddy the waters by having a professor running around promoting vaccine diplomacy in Tehran.

The next round of discussions regarding country selection were more productive. I learned about the concerns for the North African countries of Morocco and Tunisia. Youth unemployment and underemployment for both nations was high, and there were worries about creating a new generation of disaffected young people. As a result, young people in Morocco and Tunisia were vulnerable to recruitment by the Islamic State and the breeding of terrorism in the region. For example, one estimate finds that more than 1,000 Moroccans joined the Islamic State in Syria and Iraq between 2012 and 2014, and were responsible for terrorist attacks in Europe [14]. Thousands of young Tunisians also joined the Islamic State [15]. Therefore, one potential role I had as US science envoy was to build biotechnology in North Africa as a means to help create a biotech ecosystem as a potential source of new and high-level employment and economic diversification.

Another major country of interest was the Kingdom of Saudi Arabia. There, the strategic interest of the United States was the fact that Saudi Arabia was in many respects the true center of the Islamic world, certainly at least the Sunni Islamic world. In contrast, Iran as a leader of the Shia world had become a strong and bitter rival of the Saudis, which was leading to escalations in conflict in the region. For example, many Middle East experts consider the ongoing hostilities in Yemen as essentially a surrogate Sunni-Shia conflict and competition for hegemony in the region [16]. The discussions between the United States and Iran were viewed by some leaders in Saudi Arabia as a potential threat.

The State Department considered my role as science envoy in Saudi Arabia especially important because it had the potential to reduce tensions with the United States that were growing out of the new American interest in Iran. Given the fact that Saudi Arabia had some of the highest ranked research universities and institutes in the Middle East, and sophistication in the area of biotech, even if it was not necessarily in the vaccine space, I felt there was a lot to work with in terms of vaccine development. Fueling this opportunity was the fact that Saudi Arabia was sandwiched between two war zones.

I wound up making multiple trips to the region, including five to Saudi Arabia, two to Morocco, and one to Tunisia. In so doing, I became one of the few science envoys to serve two terms. For me it was a life-changing experience because it allowed me to freely explore and implement my aspirations for vaccine diplomacy. My role as US science envoy also gave me a deep appreciation for my State Department colleagues. Although they do not always have a high profile, I consider them some of the most sophisticated and committed public servants our nation has ever seen. They are true heroes.

4:

Battling Diseases of the Anthropocene

As US science envoy, I focused my energies on building vaccine capacity and joint US vaccine development with Muslim-majority nations in the Middle East and North Africa. I also hoped to explore developing new vaccines to combat diseases arising from the conflict zones linked to the Islamic State and the war in Yemen. While the collapse of public health infrastructures and systems from war became the major driver of disease in the Middle East, it was not the only promoter. The human diaspora from the Islamic State introduced new infections into neighboring Jordan, Turkey, Lebanon, and Egypt [1]. In parallel, the region now experiences unprecedented high temperatures, sometimes regularly reaching 50°C for long periods, together with floods and droughts [2]. The warming temperatures expand insect habitats and areas vulnerable to the diseases transmitted by insect vectors, while new flooding might support snail intermediate hosts that transmit schistosomiasis. Moreover, unprecedented droughts and high temperatures force human populations to abandon agricultural lands and flee into urban centers such as Aleppo in Syria. Urbanization, in turn, overburdens already fragile cities in terms of their ability to ensure food security and provide access to safe and potable water. Cities become vulnerable to outbreaks of cholera and other causes of infectious diarrhea, and ultimately COVID-19. This combination of forces—war and conflict, human migrations, climate change, urbanization, and other twenty-first-century drivers—combined in a perfect storm for the emergence of infectious diseases. Is there an overarching theme to these shifts?

The Anthropocene

In 2016, I spoke in Waco, Texas, at Baylor University's STEM and Humanities Symposium. This annual event works to bridge humanities and STEM (science, technology, engineering, and math) fields, in order to stimulate conversation and collaboration. The symposium also aspires to produce innovations and changes in university curricula. I welcomed this opportunity, because increasingly I recognized how solving complex global health problems may require us to go beyond the traditional biomedical model and begin accommodating the social and physical determinants highlighted above. Too often universities operate in silos, so that virologists or vaccinologists seldom speak to economists, political scientists, or scholars of the humanities. Government agencies also tend to work in silos. The theme of the 2016 symposium was the Anthropocene, a term with which I was only vaguely familiar at the time. But being invited to speak on this topic afforded me an opportunity to investigate it further, and I found it a useful construct for organizing the modern forces that might require vaccine diplomacy.

In 2000, the Dutch Nobel laureate and atmospheric chemist Paul Crutzen, who helped to form the concept of nuclear winter, coined the term "Anthropocene" at a conference in Mexico. The context of the Anthropocene is the contention that our human species is entering its first new geological epoch since the end of the ice age when the Holocene began, roughly 12,000 years ago [3–5]. The major argument for the Anthropocene relies on geological evidence indicating that humans have so profoundly changed our planet that we can now mark our time as a distinct epoch.

While the idea is provocative and interesting, many geologists and earth scientists remain skeptical that the Anthropocene actually exists, but some relevant findings support the theory. For example, reports from the British Geological Survey point to increases in soil phosphorous and nitrogen levels as a result of fertilizer and expanded human agricultural activity, or elevations of lead levels in the soil as a consequence of World War II, or the first appearance of radionuclides that began with atomic bomb and hydrogen bomb testing [4, 5]. There are also the rises in atmospheric carbon dioxide and methane, increased levels of which are producing climate change [4, 5]. Each of these findings represents so-called geochemical signatures of widespread human activities and their impact on the planet. Each also reflects the human activities that might drive disease introductions or spread, including climate change, but also urbanization and war.

Even among scientists who accept the Anthropocene concept, there are disagreements as to when this new epoch actually began. Some say it started with development of human civilization resulting from agriculture, while others point to the Industrial Revolution or the nuclear age [3–5]. Another group links it to an uptick in global levels of concrete and plastic, together with losses of many species of animals and plants [3–5]. I find the Anthropocene useful as an umbrella term to embrace the range of human changes to the environment now promoting the emerging and neglected diseases [5]. "Planetary health" is a related term used also to describe public health in the context of shifting ecosystems and changes to civilization. A joint Rockefeller Foundation and Lancet Commission on Planetary Health headed efforts to establish the definition and framework of this concept [6], leading ultimately to the Planetary Health Alliance, based at Harvard University [7]. Many of the forces originating in the Anthropocene that are now promoting endemic and epidemic infections are social determinants, with poverty often dominating, but increasingly we are noticing how war or conflict and political destabilization promote disease, together with human migrations and urbanization, which often go hand-in-hand. In addition, urbanization can be linked to the physical determinants of climate change—as droughts expand, temperatures rise, and populations abandon their agricultural lands and crowd into cities. Let us look at some of these individual factors in more detail.

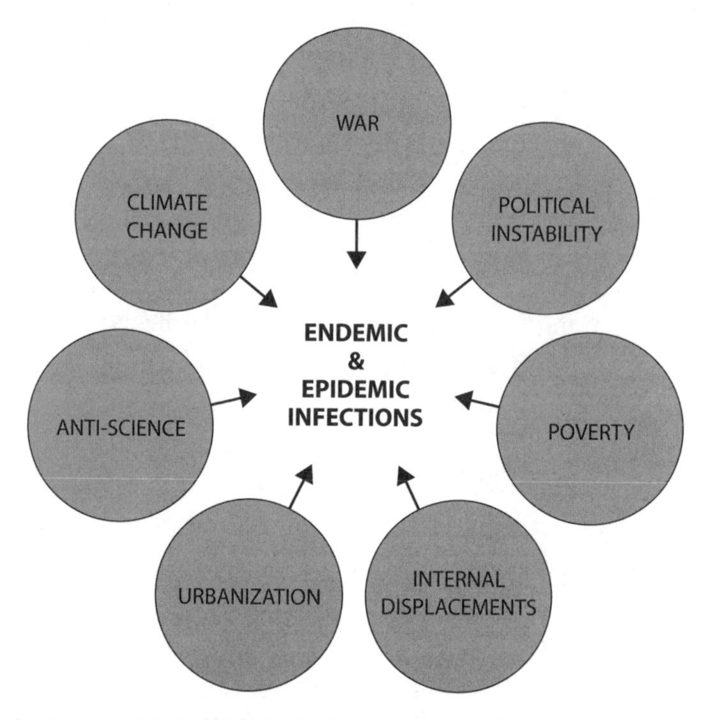

Major forces arising out of the Anthropocene that are now promoting the emergence of vaccine-preventable infections and neglected tropical diseases. Original diagram by Kenny Cuevas.

War and Political Collapse

Throughout modern history, war and political unrest dominate as drivers of epidemics of human infectious diseases. War also propagates and enhances the diseases already circulating among a population. During the twentieth century, pandemic influenza spread as a result of large movements of troops and other populations toward the end of World War I, while both World Wars produced epidemics of typhus. A key lesson from twenty-first-century geopolitics is that wherever war and extreme political instability occur, both emerging and neglected diseases follow. Cholera arose in the aftermath of the Haiti earthquake and the war in Yemen, while Ebola emerged after years of atrocities and conflict in western Africa (Guinea, Liberia, and Sierra Leone) and more recently in DR Congo. Malaria and other vector-borne diseases, as well as measles, now flourish in Venezuela following its economic collapse and in the Northern Triangle of Central America with the rise of drug cartels. The Middle East region is a new global hot zone of emerging and neglected tropical diseases.

Measuring or confirming the links between disease and conflict is not always easy, because war zones are, almost by definition, unsafe places for epidemiologists and health scientists. So conducting detailed epidemiologic investigations in times of war is seldom a realistic prospect. However, Rebecca Du, Jeffrey Stanaway, and I compared a Global Peace Index (released by the Institute for Economics and Peace) with countries now exhibiting the highest estimated prevalence rates of most of the NTDs. We found that 15 of the 18 different types of NTDs we looked at were endemic in the countries with global peace indices considered "low" or "very low" [8]. Moreover, several nations with the highest global prevalence of a particular NTD were outright engaged in war or conflict. They included Afghanistan, with the world's highest prevalence rate of cutaneous leishmaniasis; Central African Republic (the highest prevalence rates of human African trypanosomiasis); Myanmar (rabies); and South Sudan, with the highest prevalence rates of two NTDs—visceral leishmaniasis and leprosy [8]. Other endemic nations on the list were those that are considered post-conflict and still emerging from terrible atrocities, including Angola, Timor-Leste, and Liberia [8]. Chris Beyrer, a professor at the Johns Hopkins Bloomberg School of Public Health, has traced conflict to predisposing conditions that might promote disease, such as food insecurity and malnutrition or other factors that increase disease susceptibility, and to enhanced exposure to disease, especially to insect vectors that transmit viruses or parasites [9].

Among the factors linking war and conflict to disease are some critically important social determinants. They include both the destruction of healthcare facilities and the collapse of public health services (especially vaccination and mass treatment campaigns), as well as interruptions in food security systems, animal control, sewage treatment and sanitation systems, or programs for vector control to reduce the spread of disease from insects [8]. In addition, government services related to healthcare are diverted to combat, resulting in the loss of key personnel needed to support local health systems, such as doctors, nurses, and public health experts. Another factor is direct damage to the environment from war, leading to deforestation and the contamination of water, soil, and food [8]. The collapse of water and food security results in diarrheal disease and increased human susceptibility because of malnutrition. Yet another major contributor to the war and disease connection are forced human migrations and population displacements. This activity deprives populations of adequate shelter, while promoting both malnutrition and disease exposure to vectors and other disease transmission agents [8]. Displacements also deprive populations of healthcare access [8]. There is also the risk that refugee movements will introduce new diseases to a region. Adding to this mix is vio-

Factors Linking War, Conflict, and Disease

- Destruction of healthcare facilities
- Collapse of health systems and public health services
- Loss of healthcare and public health personnel
- Loss of food and water security and animal control
- Environmental damage
- Population displacements and human migrations
- Loss of government capacity
- Violence
- Increased poverty

lence, including sexual violence, and extreme exacerbation of profound poverty, itself a social determinant.

Climate Change

There is overwhelming scientific evidence that the earth is warming, most likely because of human activity [10]. Among the evidence for warming are global increases in average surface temperature of almost 1°C since the late 1800s, together with warming oceans, diminishing polar ice sheets and snow cover, rises in sea level, melting sea ice and retreating glaciers, and an increase in the number of extreme weather events [10].

In the Middle East, several cities have just recorded regular temperatures exceeding 50°C (122°F) [11], and some areas will soon become uninhabitable as a result of desertification from droughts and prolonged heat waves [12]. The Middle East droughts related to climate change may have driven rural farmers into the Syrian cities of Aleppo and Damascus, creating conditions of crowding that further promote political and civil unrest [12]. In parallel, warming temperatures and droughts drying up the Tigris and Euphrates Rivers in Iraq, and other bodies of fresh water, exacerbate intense competition and confrontations over dwindling water supplies [12].

In sub-Saharan Africa, the area within 15 degrees north or south of the equator, which includes the major conflict zones of DR Congo, South Sudan, and Central African Republic, also experiences prolonged heat waves and an increase in hot nights [13]. Moreover, central Africa will see increases in rainfall, whereas droughts will intensify in the Sahel [13]. Finally, Venezuela is experiencing its worst drought in four decades, and there are ongoing water shortages. Between the years 2013 and 2016, Venezuela experienced a 50–65% decrease in rainfall, which in turn depleted the national capacity for generating hydroelectric power and produced prolonged periods of blackouts [14]. Decreasing water reserves and drought combine to promote food insecurity, owing to decreased production of coffee, corn, and rice, as well as shrinkage of livestock herds [14].

An important conclusion is that climate change, through hotter temperatures, drought, and desertification, fosters increased insecurities in food and water in areas already beset by conflict. Climate change also drives human populations into increasingly crowded cities, with resultant breakdowns in municipal infrastructures that are meant to provide access to sanitation and safe water. Therefore, the fact that climate change affects major conflict zones might be more than just a coincidence. Instead, climate change itself might further destabilize economies and governments.

I have had the opportunity to meet and listen to the former US vice president Al Gore twice. On both occasions, he eloquently pointed out links between climate change and poverty. It disproportionately affects the poor because they are especially vulnerable to food and water insecurity, as well as to extreme weather events, given their inadequate housing and the fact they often must live in low-elevation, flood-prone areas [15].

Through these mechanisms climate change drives unregulated urbanization and reinforces both poverty and political instability. In so doing, climate change will simultaneously enhance the emergence of tropical and infectious disease.

Beyond operating indirectly through political instability and conflict, there is increasing evidence that climate change also exerts direct effects to promote the emergence of infectious and tropical diseases. A. J. Blum and I reviewed the effects of climate change on rising helminth (parasitic worm) infections. Overall, we found evidence for a mixed picture that sometimes results in conflicting or unintuitive effects. For example, intestinal worm infections affect hundreds of millions of children in poor countries, and we even recently found some evidence for intestinal worm infections in Alabama [16]. The WHO designates these diseases as soil-transmitted helminth infections because humans acquire these diseases through contact with soil contaminated with either eggs or larval stages of the parasites. One of the most common is human hookworm infection. Scientists at the London School of Hygiene and Tropical Medicine found that hookworm larvae living in African soils tolerate excessively high land surface temperatures of 40°C or more [16]. Therefore, with global warming in Africa we might expect hookworm infection to emerge as the overwhelming or dominant intestinal helminth on the African continent. As our Texas Children's Center for Vaccine Development advances a new human hookworm vaccine through clinical trials, the vaccine might find its greatest use in Africa in the coming decades.

Still another dominant parasitic helminth infection in Africa is schistosomiasis. The female schistosome residing in human blood vessels releases eggs, which exit the body in urine or feces and then hatch in water. The released larval stages enter snails living in and near large bodies of fresh water. In Africa, large bodies of water such as Lakes Malawi and Victoria teem with millions of snails containing the larval schistosomes. A similar situation occurs in bodies of fresh water found in eastern Brazil and the Middle East. The schistosome's dependence on snails provides potential links with climate change. However, warming temperatures result in conflicting effects on the prevalence of human schistosomiasis. For instance, one of the species of snails that transmits the intestinal and liver form of schistosomiasis cannot tolerate high temperatures, so climate change and warming may actually decrease the prevalence of this disease in the coming decade [16]. Offsetting these declines, however, is the possibility that some areas of southern Africa currently considered too cold to support schistosome-transmitting snails may ultimately warm up. This effect might explain why schistosomiasis transmission has recently been documented in Corsica, off the coast of France. For the first time, southern Europe now hosts a major neglected tropical disease previously restricted to the African continent and the Middle East [17].

Warming temperatures and altered rainfall patterns also affect mosquitoes and various arthropods that transmit disease. Higher temperatures can increase both the rate of virus proliferation in mosquitoes and the insect's actual development [18]. Biting and human contact rates can also increase [18]. On this basis, several groups determined that mosquito-transmitted virus infections, like snail-transmitted infections, will expand into further reaches of Africa and Europe.

In 2019, Professor Simon Hay and his colleagues at the Institute for Health Metrics and Evaluation at the University of Washington (Seattle) published their model that estimates new trends for human dengue virus infection transmitted by *Aedes* mosquitoes [19]. They predict that the global risk of dengue will expand significantly by 2050. By that time in the Americas, dengue will expand into the southern United States beyond its current focal transmission patterns in southern Florida and southern Texas, and it will reach or extend into areas of central Mexico where higher altitudes were previously associated with cooler temperatures unsuitable for *Aedes* mosquitoes [19]. For similar reasons, northern Argentina will be at risk. Moreover, some models predict that dengue will become an important public health threat in the large coastal cities of Japan and eastern China, as well as inland Australia [20]. Sub-Saharan Africa is especially vulnerable to expanded geographic ranges of dengue, especially in southern Africa and in western parts of the Sahel, which now reports only limited or sporadic cases [20]. Alternatively, in some areas dengue may actually decline, because warming temperatures might exceed the thermal limits for the survival of the *Aedes* mosquitoes. This scenario may occur in parts of eastern Africa and in India, where the rise in temperatures will be especially extreme [20]. Therefore, dengue, like schistosomiasis, will redistribute its geography. It will be interesting to see if these shifting ranges of dengue virus infection might also apply to other *Aedes*-transmitted viruses such as chikungunya or Zika virus infection.

The mother of all tropical diseases in Africa, in terms of total numbers of deaths and permanent disabilities, is currently malaria, and most of the global malaria deaths result from infection by a single species of the disease-carrying parasite, *Plasmodium falciparum*. There is no clear consensus regarding the future effects of global warming on malaria. A recent paper by Madeleine Thomson and her colleagues from Columbia University's Earth Institute highlights considerable variations in African rainfall and temperatures in the coming decades [21]. Therefore, creating models to predict how these future variations and changes might affect the rise or fall of malaria turns out to be a very complex undertaking. Among the trends noted to date are the decreases in rainfall in the Sahel toward the end of the twentieth century, causing a "retreat" of malaria there, but a more recent return of rains in the nation of Niger and elsewhere could lead to the Sahelian reemergence of malaria. Conversely, hot and drier conditions forecast for eastern Africa in the coming decades might produce declines in malaria, but some models project downstream increases in rainfall and malaria's return later in this century [21]. Warming temperatures in Africa may eventually mostly promote malaria in eastern and southern African highland regions, where cooler temperatures currently reduce the threat.

While global trends in terms of a warming planet and rising sea levels are fairly evident and predictable, exactly how climate change will specifically affect tropical disease patterns remains a work in progress. Overall, warming temperatures indicate that tropical infections, especially some soil-transmitted helminth infections, and those transmitted by insect or snail vectors of disease, will increase in prevalence in areas currently considered too cool to facilitate transmission. This observation especially seems to be the case for areas of high elevation surrounded by low-lying endemic regions. With warming, temperatures will increase at high altitudes to promote disease there. Since adequate moisture is also required to sustain disease vectors, warming temperatures would ideally be linked with increased rainfall in order to promote disease transmission. However, this is not always the case, so hot, dry, and arid conditions might actually reduce local disease endemicity.

Finally, the emergence or retreat of tropical diseases attributable to warming and changes in rainfall is not happening in isolation. NTD mass treatment programs are under way for several parasitic worm infections, so future epidemics and endemicity for these conditions might strike a balance between each disease arising in warming and wetter regions versus the retreat of disease in areas specifically targeted for mass treatment or those that will experience hot and dry conditions. The same set of circumstances might be true for the war on malaria that employs antimalarial drugs and bed nets. Still another unknown is how quickly we can accelerate the development of new vaccines for neglected and emerging diseases, so that disease control might potentially distill down to a race between vaccine development and vaccination programs and climate change.

Urbanization

It is just a matter of time before most humans on our planet will live in cities. According to the UN, within the past decade, we just surpassed the point at which the majority of the human population now lives in urban areas rather than rural ones, while some new projections indicate that by 2050 roughly two-thirds will live in urban areas [22]. Further projections suggest that African and Asian cities will show the greatest expansion. Much of the urban growth will occur in just three countries—China, India, and Nigeria [22, 23]. Africa and Asia will also show the fastest declines in rural populations.

The most direct consequence of aggressive urbanization will be the unprecedented creation of new megacities. There are different definitions of what constitutes a megacity, but most of them refer to enormous metropolises with populations exceeding 10 million people. In the coming decade or by 2030, at least 40 global megacities will emerge, and increasingly they arise in low- and middle-income nations of Africa, Asia, and Latin America. Several Chinese and Indian urban areas have already attained megacity status, but increasingly we will find them in Africa. Projected African megacities include Kinshasa (DR Congo) and Lagos (Nigeria). Regarding these two cities, some projections from the Gates Foundation indicate that by 2050 approximately 40% of the world's poor will live in DR Congo and Nigeria [24, 25]. We might anticipate that many of those people living in extreme poverty will emigrate from rural areas and crowd into Kinshasa and Lagos, respectively. Moreover, with climate change disproportionately affecting the poor and those in resource-poor countries, we should anticipate that the megacities of Africa might soon experience an expanding "triple threat" of aggressive urbanization, extreme poverty, and climate change.

Without additional or new interventions, megacities such as Kinshasa and Lagos are at the highest risk for succumbing to that triple threat, as they outstrip the urban infrastructures required to sustain populations in adequate health. Rivers and other sources of fresh water will be contaminated by pollutants and infectious agents; sanitation systems will collapse; and housing will become inadequate and of universally low quality, lacking plumbing, screens, and air conditioning. Crowding will become intense. Some might argue that these events have already taken place. Food insecurity might also become rampant.

I am also concerned that the pathogens themselves, or their vectors, could become increasingly adapted to an urbanized environment. For example, we are now seeing more and more reports of urban or peri-urban transmission of parasitic worm infections, including soil-transmitted helminth infections and schistosomiasis [22]. What remains unclear is whether the developmental stages of these parasitic worms, typically eggs or larvae, have somehow evolved to better adapt to urban life. There is also a parasitic worm infection found in the United States that appears to have adapted itself well to urban environments. *Toxocara* worms live in the intestines of stray dogs and cats that roam in degraded urban areas. The parasite eggs of *Toxocara* contaminate the soil where kids play, so a high percentage of children living in urban slums are at risk for becoming infected. When children accidentally swallow *Toxocara* eggs, the newly released immature or larval worms travel through the lungs or brain to cause an illness that resembles asthma in the case of the lung migrations, or cognitive and developmental delays resulting from central nervous system migrations [23, 26]. I have outlined how toxocariasis may represent an important cause of the achievement gap noted among socioeconomically disadvantaged children [23, 26]. Toxocariasis is also widely prevalent in Brazil and elsewhere in South America.

An important but unanswered question is whether urban transmission is old or new. As suggested by their "neglected" moniker, NTDs do not benefit from intensive study, so the biomedical literature on these diseases is not extensive. An intriguing but unproven hypothesis relies on parasite adaptation to an increasingly urbanized human population. As we move into cities, especially the hot, crowded, environmentally damaged, and impoverished cities of Africa, Asia, and Latin America, could it be that the pathogens are evolving to these new circumstances? Could parasites adjust to transmission in urban landscapes, just as their human hosts? This could represent an important new model.

I also predict that several serious bacterial infections will emerge when cities outgrow their capacity to provide safe water and adequate sanitation. Two of the most serious in terms of high fatality rates are typhoid fever and cholera. These diseases also commonly arise in settings of twenty-first-century conflict and political instability. The good news is that safe and effective vaccines are now available for these diseases, but overall access to them is still poor to modest. Leptospirosis is another bacterial infection thriving in some urban environments, especially in widespread areas of environmental degradation. Leptospira are unusual but often neglected bacterial pathogens that live in the kidneys of urban rats and dogs and then infect humans through their contact with water contaminated by animal urine. Two of the leading leptospirologists are Drs. Albert Ko and Joseph Vinetz at Yale University, who have studied this disease extensively in the urban slums and favelas of Latin American cities. The disease can result in fever and a life-threatening illness known as Weil's disease, characterized by high fever and jaundice (yellowing of the skin resulting from liver damage). Urban areas of Africa currently report cases of leptospirosis, so we should expect this disease to rise to prominence in Lagos, Kinshasa, and other resource-poor megacities. Canine rabies is another serious infection transmitted by dogs (in this case, a virus) and associated with a high fatality rate; it is another illness we might expect to emerge in megacities.

I also predict that the urban-dwelling mosquito, *Aedes aegypti*, will become a dominant vector in future megacities located in tropical and subtropical environments. The diseases transmitted by this insect include dengue, yellow fever, chikungunya, and Zika virus infections. We have already seen how both chikungunya and Zika virus infection rapidly spread across urban areas of the Americas beginning in 2013, with the 2016 Zika epidemic culminating in high rates of birth defects, especially microcephaly, in Latin American cities such as Recife in northeastern Brazil. We should anticipate that explosive epidemics of *Aedes aegypti*–transmitted viruses could become common on an urbanized planet Earth. My former Yale colleague Dr. Mark Wilson and his collaborators further note how malaria in some instances may now be transforming "from a rural to an urban disease" [27]. Refugees are now fleeing rural areas of conflict and pouring into cities. As this occurs, malaria increasingly could shift from rural to urban transmission patterns. In some cases, this process may involve a switch from a rural-adapted *Anopheles* mosquito species to one better suited for cities. In the future, we will look out for evidence of urban adaptation of the *Anopheles* mosquitoes that transmit malaria.

Aside from the viruses transmitted by mosquitoes, there will be a rise in respiratory virus infections. COVID-19, caused by the SARS-2 coronavirus (SARS CoV2), is a highly transmissible virus infection that spreads quickly in dense urban areas. We have seen how epidemics quickly accelerated in Wuhan and other cities of central China before they moved to other crowded urban centers in Asia, Europe, and then North America. The fact that COVID-19 caused a devastating epidemic in central Queens in New York City is not a coincidence. The interconnected Queens neighborhoods of Jackson Heights, Elmhurst, Corona, and others are teaming with working-class immigrants who live amidst a vibrant street life, but this densely populated area ultimately became the epicenter for COVID-19 in the United States. Many of us now anticipate that COVID-19 could cause major epidemics in the new megacities of Asia, Africa, and Latin America.

Along with the rise of urbanized infections and NTDs, we should also expect that diabetes, heart disease, and cancer will simultaneously become chronic noncommunicable disease conditions in resource-strapped megacities, particularly urban areas where food insecurity, poor diets, and widespread access to tobacco are significant. Priyanka Mehta and I have noted how we are increasingly seeing a confluence of epidemics of both NTDs and noncommunicable diseases, with some individuals showing signs of having these two conditions simultaneously. For example, in India, individuals who are often the sickest from dengue fever are those with underlying diabetes and hypertension [28]. The same is true in Texas for individuals infected with tuberculosis. We are now seeing a similar situation unfold during the COVID-19 epidemic in the United States, where the health disparities among African American, Hispanic, and Native American populations, which suffer from high rates of diabetes, hypertension, and obesity, are associated with more serious cases of the illness. Exactly why or how NTDs and noncommunicable diseases combine to produce severe illness is still an important but mostly uninvestigated health threat.

However, in the new world of megacities, I expect we will begin to see many more examples of comorbidities. This is also an important avenue for future study and a future disease paradigm.

Shifting Poverty

Conflict and war, urbanization, and climate change rank high as key factors that drive the emergence or reemergence of tropical infectious diseases, but poverty still remains the king of all determinants. NTDs flourish in the setting of extreme poverty and, simultaneously, NTDs reinforce poverty through several mechanisms. They include rendering adults too sick to work effectively and provide for families, while for children, NTDs impair their intellectual and cognitive development [29, 30]. Some of the most highly prevalent and widespread NTDs also disproportionately affect girls and women or occur commonly in pregnancy, increasing the risk of death or serious injury to the mother and often causing prematurity or decreased neonatal survival [31, 32].

For all of these reasons, interventions that treat or prevent neglected diseases can become potent and innovative antipoverty measures [29]. The Millennium Development Goals fully embraced these measures, and they were effective. The World Bank now estimates that the number of people living in extreme poverty has decreased from roughly 25% of the world's population (1.5 billion people) to below 10% (less than 700 million) [33]. These poverty reductions also coincided with significant declines in poverty-related diseases [34], although the extent to which poverty decreased directly because of these disease burden reductions requires additional study.

With global improvements in economic development, the geographic regions of poverty-related neglected diseases will shrink and concentrate in remaining areas where poor people concentrate. While projections from the Gates Foundation indicate that by 2050, 40% of the world's impoverished population will live in Nigeria and DR Congo, for now and in the coming decade I have noted a very different trend: that NTDs and other poverty-related neglected diseases are currently more global. Based on an analysis of data released by the WHO and the Global Burden of Disease Study, I find that today, most of the world's poverty-related diseases, including the NTDs, occur in the Group of 20 (G20) nations, together with Nigeria [35].

The 19 largest national economies, along with the European Union make up the G20. Currently, the United States has the largest economy, with a GDP that exceeds $20 trillion, followed closely by the European Union and China, and then Japan and India. The G20 itself is a relatively recent creation, and one born out of necessity stemming from financial crises. It began in 1999 to promote international cooperation in the wake of an Asian financial crisis and held its first summit during the Great Recession that began in 2008 or at the end of 2007 [35]. Together the G20 and Nigeria account for almost 90% of the global economy or gross domestic product. Why would we expect these nations to host large populations with NTDs and other poverty-related neglected diseases? I found that some of the largest G20 nations simultaneously encompass both enormous wealth and profound poverty [35]. NTDs flourish in the poverty pockets found in each of the G20 countries. In fact, the G20 nations and Nigeria actually account for most of the world's people infected by NTDs such as helminth infections, leishmaniasis, Chagas disease, dengue, and leprosy, as well as the "big three diseases"—HIV/AIDS, malaria, and tuberculosis.

I named the concept of poverty-related diseases in wealthy nations "blue marble health" in order to distinguish it from traditional norms of global health. Whereas global health focuses heavily on overseas development assistance from rich countries going to the world's poorest and most devastated nations, blue marble health recognizes the widespread NTDs among the poor living in wealthy countries, a group sometimes referred to as "the poorest of the rich" [36]. In terms of public policy, blue marble health finds that if the leaders of the G20 nations redoubled their attention to the poorest and most vulnerable populations within their borders, we could achieve dramatic reductions in the prevalence and incidence rates of NTDs. Many of these nations, such as Brazil, Mexico, and Saudi Arabia, also have tremendous potential for innovation, and we work there to develop jointly new antipoverty vaccines.

To take an example in the Western Hemisphere, the three largest economies in Latin America—Argentina, Brazil, and Mexico (each a G20 nation)—overwhelmingly host the majority of the global cases of Chagas disease, a parasitic illness of extreme poverty that leads to debilitating or fatal heart disease and is transmitted by the "kissing bugs" that infest low-quality housing. Yet well over 90% of people living with Chagas disease in Argentina, Brazil, and Mexico do not have access to diagnosis and treatment for this condition. In the United States, I estimate that at least 12 million Americans suffer from at least one NTD [35].

Another key point regarding NTDs in wealthy nations in the Americas is that they are not homogenously distributed. In Argentina, NTDs such as Chagas diseases, leishmaniasis, and helminth infections proliferate mostly in the northern part of the nation, especially the Gran Chaco region, where there are high rates of rural poverty. Similarly, poverty and disease in Brazil concentrate in the northeastern states and cities such as Recife and Salvador de Bahia, where Zika virus infection and microcephaly first emerged, but where Chagas disease, leishmaniasis, schistosomiasis, lymphatic filariasis, and other helminth infections also occur. Southern Mexico hosts the highest concentration of Chagas disease, leishmaniasis, and helminth infection. Poverty and disease often concentrate in specific regions of each of the G20 nations.

Outside of the Americas, in China, dramatic economic prosperity over the past few decades in eastern Chinese provinces has equated to substantial reductions in disease, especially in major cities such as Beijing and Shanghai. However, extreme poverty and disease remain widespread in western Chinese provinces such as Guizhou, Sichuan, and Yunnan. Conversely, eastern Europe has a significant level of parasitic and related neglected diseases not usually found in the more prosperous countries to the west. In the Northern Territory of Australia, NTDs affect aboriginal populations. The diseases include scabies and secondary impetigo bacterial superinfections, trachoma, and strongyloidiasis. Similarly, in northern Canada, indigenous Arctic populations suffer from trichinellosis, toxoplasmosis, and other neglected zoonotic infections, transmitted from animals. So far, the COVID-19 pandemic has disproportionately affected the G20 nations, led by China, the United States, and European nations.

In several cases, the G20 nations and their leaders have taken true ownership of their NTDs and worked to eliminate them within their borders. For instance, both Japan and South Korea suffered from terrible deprivations and extreme poverty in the immediate aftermath of World War II and the Korean War, respectively. In Japan, both soil-transmitted helminth infections and schistosomiasis were widespread, while soil-transmitted helminth infections were highly prevalent in both North and South Korea. Through a combination of significant economic development and mass deworming programs, both Japan and South Korea mostly eliminated many of their NTDs. In contrast, these diseases remain endemic in North Korea, as exemplified by a recent report of *Ascaris* worms recovered from a North Korean defector who suffered gunshot wounds as he tried to escape and cross the border into South Korea. Moreover, both Japan and South Korea have recently created technology funds and infrastructure to develop new drugs, diagnostics, and vaccines for neglected diseases, including the Japanese Global Health Innovative Technology and South Korean Research Investment for Global Health Technology Funds. However, most of the other G20 nations enable both poverty and NTDs, and their governments so far do not support developing new neglected disease technologies at a substantial scale. This includes several nations with the technical capacity to develop and deliver nuclear weapons [37].

Nationalism

Related to the concept of diseases of the poor in the G20 are emerging trends linked to nationalism. The establishment of the Millennium Development Goals created the framework for overseas development assistance that provided unprecedented access to vaccines and essential medicines for HIV/AIDS, tuberculosis, malaria, and NTDs. However, this flourishing of globalism may be ending. The 2016 election of President Trump and his "Make America Great Again" campaign may signal the start of a new period of nationalism marked by conservatism, transactional foreign policies, militarism, and economic protectionism [38]. The same might be said of post-Brexit England and several European nations, such as Italy and Hungary, or Brazil in South America or China and Indonesia in Asia. These nationalist trends create new challenges for G20 leaders to provide development assistance for health [38]. The "poorest of the rich" framework could provide mechanisms for nationalists to support treatments or research and development for NTDs and other poverty-related neglected diseases. Widespread NTDs now represent important drags on the economic development and prosperity of the G20 nations. Today, one of the most cost-effective mechanisms for accelerating the economies of just about all of the G20 nations would be to reduce the disease burden from the NTDs and related illnesses. Even nationalist regimes should see the benefit of treating and preventing their NTDs or investing in antipoverty technologies.

5:
The Middle East Killing Fields

I cherish the memory of my time as US science envoy for the Middle East and North Africa. For me, it was the culmination of decades of work in developing vaccines for poverty-related neglected diseases, together with my broader outreach in science policy and public engagement (especially concerning vaccines and neglected tropical diseases). I had also written articles about the historical importance of vaccine diplomacy and was deeply involved in developing our human hookworm vaccine jointly with Brazilian scientists. More recently, we are working to develop our Chagas disease vaccine with a consortium of Mexican institutions and, now, a new COVID-19 vaccine with India.

However, up until the time of my science envoy appointment, I did not really have the opportunity to pursue vaccine diplomacy as Albert Sabin or D. A. Henderson did with the Soviets during the 1950s or 1960s. For me, working on the problem of vaccine science and development in or near conflict areas of the Middle East or North Africa and at the height of the Islamic State occupation represented a true twenty-first-century equivalent.

I was exhilarated by the prospect of conducting vaccine diplomacy in the Middle East and North Africa. However, at the same time, I recognized the very real dangers of being in this part of the world in 2015. All three of the major countries where I visited and worked—Tunisia, Morocco, and Saudi Arabia—either had experienced a recent significant terrorist attack or suffered a terrorist attack during my two years as an envoy. In Tunisia in 2015, there was a mass shooting at a tourist beach resort that killed 30 British citizens. This occurred just a few months after another mass shooting at the Bardo National Museum in Tunis. Either the Islamic State or al-Qaeda carried out those attacks. In addition, Morocco suffered a serious al-Qaeda bombing in Marrakesh in 2011, and news reports claimed that Islamic State cells composed of Moroccans were operating in Libya in 2016. Saudi Arabia in 2015 experienced several terrorist attacks in the Shia minority region in the east, but there also were concerns about the Islamic State or al-Qaeda targeting members of the royal family or otherwise working to destabilize the ruling government.

Although I was designated US science envoy, I did not travel with a diplomatic passport or with a security detail, so in some respects I was a soft target. However, I was usually accompanied by someone from the US State Department (often it was Dr. Bruce Ruscio who became a good friend and colleague), or by a US embassy staff member. I remember a senior official at the embassy in Saudi Arabia once explaining to me its heavy fortifications (and marine guard) made it largely impervious to a conventional attack. He gave me sort of a shrug when I explained that I was staying in a nearby hotel outside the diplomatic compound. As much as I was excited about serving as US science envoy for this region, I was also secretly relieved when my flight returning home was wheels up.

Sunni-Shia Rivalries

Volatility in the Middle East has a number of origins. Some problems date back to the years after World War I following the collapse of the Ottoman Empire, when the European powers redrew national borders to suit their colonial interests rather than to reflect tribal and ethnic differences in the region. Then, beginning with the Iranian revolution in 1979, the Middle East destabilized because of a series of conflicts between Iran and Iraq or surrogate conflicts between Iran and Saudi Arabia. Some Middle East scholars consider the rivalries between the Shia Muslim minority centered in Iran and a Sunni Muslim majority based in Saudi Arabia as the greatest generator of regional tensions [1]. The political scientist Vali Nasr at Johns Hopkins University designates the wars in Syria, Iraq, Yemen, and elsewhere in the region as an "Iranian-Saudi rivalry by proxy" [1].

In 2014, the Sunni Islamic State of Iraq and the Levant, also known as the Islamic State of Iraq and Syria, or simply the Islamic State (or its acronym in Arabic), captured global attention when it conquered large territories by driving out the government forces of Iraq and declared itself a caliphate. Over the ensuing five years, the Islamic State conducted public executions and practiced sexual violence, atrocities, and human rights abuses on a scale not seen since World War II. At its peak, the Islamic State boasted an army of tens of thousands of fighters, while generating a large budget from captured oil reserves in Iraq and elsewhere. It also operated through affiliates in at least a dozen countries in the Middle East and North Africa.

Ultimately, by 2019, through a combination of US-led coalition air strikes named Operation Inherent Resolve, the Islamic State lost almost all of its caliphate territory in Iraq and Syria. However, during the five years of its reign of terror, thousands of people were killed, injured, or enslaved. The collapse of health systems and public health control during this period also ignited epidemics and disease emergence in Syria and Iraq. Then the ensuing human migrations promoted the spread of disease into neighboring Jordan, Lebanon, Libya, Turkey, and elsewhere.

Aside from the war injuries and psychological trauma resulting from the Islamic State occupation [2], the greatest impact has been a resurgence of infectious diseases. These include infectious complications of war injuries such as bone infections (osteomyelitis), often caused by antibiotic-resistant bacteria [3] and tuberculosis [4]. In addition, as animals are trafficked across international borders, we are seeing a resurgence of rabies and of diseases transmitted from animals to humans, also known as zoonoses [5]. Regrettably, the Islamic State occupation interrupted vaccinations, resulting in a return of vaccine-preventable diseases, including hepatitis A, polio, measles, and others [6–8]. These diseases have been difficult to contain owing to the challenges of organizing and implementing catch-up vaccination campaigns.

One of the most dramatic and notable rises of an infection stemming from the chaos and collapse of countries on the Arabian Peninsula has been the parasitic disease known as cutaneous leishmaniasis. Known as "Baghdad boil," "Aleppo evil," or "one-year sore," cutaneous leishmaniasis is a parasitic infection of the skin and underlying tissue caused by various species of a single-celled protozoan parasite of the genus *Leishmania*. The disease is transmitted through the bite of tiny, blood-feeding sand flies, now proliferating in the uncollected garbage and waste in large urban areas of the conflict zones and killing fields of the Islamic State. The sand fly inoculates *Leishmania* parasites at the site of its bite, whereupon the parasites multiply to produce a large and often disfiguring ulcer. The dramatic collapse of Syrian and Iraqi health systems was accompanied by an almost equally dramatic rise in the number of cutaneous leishmaniasis cases, roughly a 10-fold increase in Syria to more than 200,000 cases annually, with another 100,000 cases annually in Iraq [9]. Afghanistan as well has suffered from high rates of cutaneous leishmaniasis.

Cutaneous leishmaniasis is generally not a disseminated or fatal disease, and usually it is self-limited to a large skin ulcer. It self-heals after weeks or months but often only after leaving a permanent scar. A lifelong problem results when the scar occurs on the face. The resulting psychological effects from the social stigma linked to facial scarring can be devastating, especially for girls and women. In collaboration with colleagues from the Liverpool School of Tropical Medicine, WHO, and the Saudi Arabian Ministry of Health, various colleagues and I undertook studies to look more closely at the long-term complications of this malady [10, 11]. Among the more interesting findings is that current efforts to measure the public health impact and disease burden of cutaneous leishmaniasis largely ignore the effects of permanent scarring. Therefore, disease burden estimates miss almost all of the years lived with disability from this disease. For instance, recorded estimates of the number of annual cases is placed around 2–4 million, but we find as many as 40 million people may now be living with the chronic effects of the ailment, mostly in the Middle East and Central Asia, but also in South Asia, Africa, and Latin America [10]. Still another interesting finding is the high rate of psychological morbidity, ranging from some level of distress to outright clinical depression [11]. Our estimates indicate that 70% of people with cutaneous leishmaniasis suffer from some serious psychological effects [11].

Another important factor is the impact of the Syrian and Iraqi refugee crisis on the spread of cutaneous leishmaniasis across the Middle East. Approximately one-half of Syria's population of 20 million people became either internally displaced or fled the country into Egypt, Jordan, Lebanon, or Turkey. The consequence of this mass migration is the potential spread of cutaneous leishmaniasis into these countries, where sand flies are abundant [12]. For instance, the Syrian refugee crisis has generated a new outbreak of cutaneous leishmaniasis in Lebanon [13].

The Collapse of Yemen

The disease and conflict situation may even be worse in Yemen. In modern times, Yemen has ranked among the poorest countries outside of sub-Saharan Africa and the poorest nation in the Middle East. More than one-third of its population lives in poverty, while approximately one-half of the population is at risk for food insecurity and malnutrition [14]. Many of these deprivations are the consequence of political instability and internal conflicts that worsened in 2010 and culminated in a full-blown war beginning in 2015 between a Houthi tribe armed rebel movement and a Yemeni government led by Abdrabbuh Mansur Hadi. Both the Islamic State and al-Qaeda operate in Yemen, thereby further complicating the conflict. Also fueling the tensions and the conflict is military and tactical support from bitter rivals—Saudi Arabia and Iran—each backing opposite sides. Saudi-led airstrikes have been particularly destructive in terms of human life lost and the collapse of health system infrastructures.

The civilian casualty toll from the Yemeni conflict has been devastating. Official estimates indicate that 8,757 deaths have occurred among civilians since March 2015, when the latest war began, until the end of 2018, while more than 50,000 were injured over this period [15]. Furthermore, over 3 million people have been displaced; health and agricultural systems have collapsed; and more than one-half of the population (at least 16 million people out of a population of 30 million) now lacks access to basic sanitary services and clean water [15].

The results of the scarcity of clean water and food are predictable. Famine is now widespread, as is severe stunting of children's growth as a consequence of long-standing malnutrition [14]. Compounding an acute level of childhood malnutrition is widespread diarrheal disease, including cholera [15, 16]. According to some reports, Yemen is now experiencing one of the largest cholera epidemics in "recorded history" [15, 16]. There have been more than one million cases since the last quarter of 2016, with more than 2,000 deaths [15]. Roughly one-half of Yemen's cholera cases have occurred among children, many of whom were already stunted and weakened from chronic malnutrition [15, 16].

An additional problem adding to the misery of Yemen's cholera epidemic is the observation that stockpiles of an oral cholera vaccine, which might have prevented many deaths, were not deployed in this epidemic until 16 months after it began [16]. In 2010, Harvard University's Matthew Waldor, John Clemens (who now heads an international diarrheal disease center in Bangladesh), and I called for the US government to begin stockpiling cholera vaccine in order to shorten the time frame for its deployment following large-scale outbreaks [17]. Instead, in 2013, the WHO established such a stockpile comprising an estimated two million oral cholera vaccine doses through financing from Gavi, the Vaccine Alliance, so vaccine access now plays a vital role in combating the epidemic in Yemen [18]. Some experts believe that climate patterns related to dry and wet seasons in Yemen may also exert important effects in helping to promote Yemen's cholera epidemic [15].

The four years of the conflict in Yemen have produced a disastrous disease situation as bad as, or perhaps even worse, than in the Islamic State–occupied areas of Syria and Iraq [19]. There have been international calls for Saudi Arabia to halt airstrikes and begin peace negotiations, possibly with Iran, but so far, this humanitarian disaster continues unabated. However, there are some new reports that the COVID-19 pandemic may have the unintended consequence of helping to effect a temporary cease-fire. Undoubtedly this situation partly reflects the devastation of COVID-19 on the population of Iran—now one of the top 10 nations in terms of confirmed cases—as well as Saudi Arabia.

The full death toll resulting from the conflict is still not known, but according to Save the Children, up to 85,000 children under the age of five may have perished between the start of the conflict in 2015 and the end of 2018, as a direct consequence of war, famine and starvation, and disease [19].

Hajj and Umrah

Besides the wars in Syria, Iraq, and Yemen, another potent force is promoting infectious diseases on the Arabian Peninsula. Every year the Hajj brings two- to three million pilgrims from across the Muslim world to the holy city of Mecca, while the Umrah represents a similar pilgrimage conducted at different times of the year but peaking during the Ramadan fasting month. The largest numbers of Muslim pilgrims come from some of the most NTD-endemic countries globally. Annually, they include more than 100,000 immigrants each from Indonesia, Pakistan, India, and Bangladesh, and almost as many from Nigeria and Egypt [20]. Today, most of the Muslim pilgrims transit through the airport at Jeddah on their way to Mecca.

I have traveled through Jeddah near the time when Saudi Arabia was preparing for the Hajj. The Saudis have set up a vast Hajj terminal structure of fiberglass and fabric resembling a sort of New Age tent city. It is impressive in its design, particularly because of its ability to reflect the hot sun and provide some level of cooling and shade for visitors. At the same time, the Hajj terminal creates a vast mixing bowl for infectious diseases to circulate, and serious epidemics of respiratory viral and bacterial infections often occur. Appropriately, Hajj and Umrah visitors are required to provide a certificate attesting to their vaccination status for meningococcal disease. The Saudis also recommend influenza vaccinations and certification for polio vaccination depending on the country of origin.

The Hajj Terminal at King Abdulaziz International Airport in Jeddah, Saudi Arabia. Photo by Wikimedia Commons user Shah134pk, https://creativecommons.org/licenses/by-sa/4.0/deed.en.

Another concern is that many of the Organisation of Islamic Cooperation nations are endemic for dengue and other mosquito-borne virus infections, so the Hajj and Umrah potentially facilitate disease introduction into Saudi Arabia and possible subsequent spread across the Middle East. The *Aedes aegypti* mosquito is an urban-dwelling insect that has gained a foothold in the western part of the Arabian Peninsula, including in and around Mecca and Jeddah. This mosquito is capable of transmitting dengue, Zika virus, chikungunya, yellow fever, and Japanese encephalitis. The risk is that if one or more of these mosquito-transmitted viruses are introduced by the Hajj or Umrah, it will be taken up by the local *Aedes* mosquitoes. Dengue was introduced through Jeddah into Saudi Arabia in 1994 and then again in 1997 [20]. It is for that reason that the Saudi government requires proof of vaccination against yellow fever in visitors from nations where it is still endemic. Malaria may also have been introduced via the Hajj or Umrah, as have tick-borne viruses such as Crimean-Congo hemorrhagic fever, cholera outbreaks, and possibly other diseases [20]. In parallel, because climate change disproportionately affects the Middle East, these effects combine to promote the survival of mosquitoes and other insect vectors, with an uptick in the spread of dengue and other vector-borne diseases [21].

Southern Europe

Could the diseases now emerging in the Middle East and North Africa also spread to Europe? Southern Europe now suffers from sharp increases in vector-borne diseases or diseases transmitted by snails [22]. Malaria has returned to Greece and Italy decades after its elimination, while arbovirus infections such as West Nile virus, chikungunya, and dengue have also emerged in Italy, Spain, and Portugal. Schistosomiasis, a parasitic worm infection transmitted by snails, was detected for the first time on the island of Corsica, the birthplace of Napoleon Bonaparte off the coast of France. It is still unknown whether refugee movements coming across the Mediterranean from North Africa could account for some of these diseases, or whether the warming temperatures in southern Europe, as well as economic downturns in countries like Italy and Greece represent the major drivers. To add to southern Europe's misery is the fact that in the spring of 2020 this region emerged as an epicenter of the COVID-19 pandemic, especially in Italy and Spain.

Middle East Vaccine Diplomacy

The Arabian Peninsula has become a global hot zone for the rise of global infections. As US science envoy, it became clear to me that the rise of disease—resulting from the confluence of war, human migrations, and climate change—was practically inevitable. I had a unique opportunity to report on my concerns to the leadership and key ministers of the Saudi Arabian government. I explained how to their north in Syria and Iraq, leishmaniasis and vaccine-preventable diseases were returning. To their south in Yemen, cholera and vaccine-preventable diseases were widespread. In the Kingdom of Saudi Arabia itself, infectious diseases, including vector-borne diseases, were being introduced on a yearly basis through the Hajj and Umrah.

In response, the Saudi government established a new center for NTDs headed by my friend and colleague Dr. Waleed S. Al Salem, who trained at the Liverpool School of Tropical Medicine. However, I felt there was an urgent need to develop countermeasures for the new diseases in the region, especially new vaccines, but there were minimal capabilities outside of some activities in Iran, as noted previously [23]. Also, at that time, neither the Saudis nor my colleagues within US government showed much appetite for negotiating vaccine diplomacy with Iran. In addition, the Institut Pasteur in Tunisia produces Bacille Calmette-Guérin (BCG) vaccine for tuberculosis and exports some its stockpile to Algeria, but it is unclear if this operation could expand significantly to meet the large vaccine demands of the region. The major multinational vaccine manufacturers have so far shown little interest in developing new vaccines of regional importance to the Middle East. For all these reasons, we embarked on a new approach to expand our Texas Children's Center for Vaccine Development to create vaccines for diseases arising from the wars, health system collapses, and political instabilities of the Arabian Peninsula. Such vaccines represented technologies considered vital to the health security of Saudi Arabia and neighboring Gulf countries.

In 2015, together with the US ambassador to Saudi Arabia, Hon. Joseph Westphal, I signed a unique agreement to collaborate with the Saudis for a joint vaccine science partnership, focused on the diseases arising from the political instability and other forces in the region. The signing ceremony in Riyadh was an exciting event for me, because it was a concrete benchmark for vaccine diplomacy. Shortly afterward, I can remember distinctly the sounds of the muezzin for the *adhan* (call to prayer), and thinking deeply about what just transpired. I was especially proud to present a new face of vaccine diplomacy for America.

6: Africa's "Un-Wars"

War and conflict are as pervasive in sub-Saharan Africa as on the Arabian Peninsula. According to some estimates, Africa experiences more than one-half of the world's global conflicts despite accounting for only about one-sixth of the global population [1]. Political scientists can identify about a half-dozen areas beset by serious conflict, despite the fact that overall peace is on the rise across most of the African continent. The conflict areas tend to cluster in major hot spots, including northern Nigeria, where the violence is linked to Boko Haram—the Islamic State in West Africa; in central and eastern Africa around the DR Congo, Central African Republic, Burundi, and Southern Sudan; and further east, where a terrible Somali civil war rages in eastern Africa [1]. There is also a highly unstable area in western Africa centered in Mali [1].

In these four major areas of conflict and post-conflict Africa, we are seeing an increase in the most common NTDs, including those diseases we first identified in the early 2000s as targets for a package of medicines and preventive chemotherapy [2, 3]. Whereas the number of new cases of these NTDs is elsewhere mostly declining, especially lymphatic filariasis, onchocerciasis, and trachoma, in Africa's conflict zones the incidence of the major NTDs is static or even increasing. Superimposed on this plateau or even rise in NTDs, is the finding that some of the same conflict-associated diseases we previously identified in the Middle East are also abruptly increasing in sub-Saharan Africa. These NTDs include cholera and leishmaniasis—but a different form of leishmaniasis, known as visceral leishmaniasis or "kala-azar." In addition, the last remaining cases of human African trypanosomiasis (also known as African sleeping sickness) are now occurring in conflict or post-conflict DR Congo, South Sudan, and Central African Republic.

Non-state actors, including local rebel groups, lead much of the violence and conflicts in Africa, but not exclusively. Increasingly, the actual governments themselves may be working behind the scenes and be directly or indirectly involved in war-related atrocities [4]. In other words, there is also a high level of hidden state-based violence. The involvement of weak or corrupt governments accounts for some of the conflict leading to disease in Nigeria, Mali, Somalia, Central African Republic, and South Sudan [4].

The journalist and former East African bureau chief of the *New York Times*, Jeffrey Gettleman, further points out that Africa's worst and bloodiest conflicts and atrocities arise and never seem to end because they are not based on ideologies or goals but instead represent what he calls "un-wars." Un-wars are mostly opportunistic enterprises, focused on what he terms "heavily armed banditry" [5]. Gettleman notes that most of Africa's combat consists of soldiers waging war on civilian populations, rather than soldiers versus soldiers. He further observes that this current situation represents a significant change, even from merely a decade ago, when rebel leaders or leaders of armed militias, many of whom are still revered, maintained an ideology. For instance, he explains that whereas Robert Mugabe of Zimbabwe was once a "guerilla with a plan" or that John Garang of South Sudan led a liberation army, such a paradigm no longer applies [5]. In its place, the civilian atrocities increasingly resemble those we identified in the Islamic State. In both Africa and the Middle East, soldiers are more like predators, and terror itself is often an end goal rather than a means to an end [5].

Some of Africa's greatest atrocities and un-wars now occur in remote, rural settings where NTDs are already widespread and that are often too out-of-the-way for access to essential medicines. In 2018, a Stanford University group of scientists and policy experts led by Eran Bendavid conducted a study looking at armed conflict and child mortality in Africa [6]. Their analysis examined more than 15,000 conflict events resulting in almost one million combat deaths over a 20-year period between 1995 and 2015. They found that the risk of an infant dying within 50 km of an armed conflict was far higher than for infants living outside of conflict areas in Africa. Moreover, the number of infant deaths was three to four times higher than the "direct deaths from armed conflict." Infants were dying in conflict zones as a result of hunger, malnutrition, and famine. The bottom line is that children in Africa mostly die in conflict zones, not from bullets and weapons, but from the indirect effects of armed hostilities [6]. Moreover, the numbers are not small—millions of children in Africa now perish under these circumstances.

Nigeria and Boko Haram

Boko Haram is the name given to the Islamic State in West Africa. Located primarily in northern Nigeria and extending into Cameroon, Chad, and Niger, the group originally formed as a nonviolent Islamic organization but later adopted the ideologies and practices of the Islamic State. Boko Haram is responsible for the murder of tens of thousands of individuals, while practicing sexual violence on an unprecedented scale and displacing more than two million people.

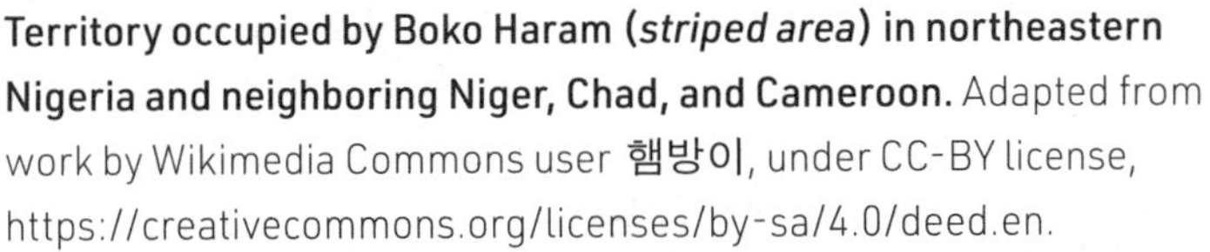

Territory occupied by Boko Haram (*striped area*) in northeastern Nigeria and neighboring Niger, Chad, and Cameroon. Adapted from work by Wikimedia Commons user 햄방이, under CC-BY license, https://creativecommons.org/licenses/by-sa/4.0/deed.en.

Boko Haram has destabilized an area of Nigeria and neighboring countries that was already a major hot zone of NTDs. Nigeria at one time was labeled "ground zero" for these diseases because as a nation it leads in so many NTD prevalence categories [7]. Now this situation has been greatly exacerbated by the Boko Haram's reign of terror, causing the collapse of health systems and the interruption or destruction of key telecommunications networks, which have largely cut off the region from modern communication [8]. With respect to infectious and tropical diseases, just as in the Islamic State–occupied areas of the Middle East and North Africa, we are now seeing the return of vaccine preventable-diseases and a rise of NTDs, including cholera.

In 2016, poliovirus infections were detected in Boko Haram–occupied areas, with genetic evidence indicating that polio transmission may have been occurring since 2011 [8]. Implementing mass vaccination efforts for polio and other infections has become highly challenging because of human displacements, with satellite images revealing the destruction of structures or facilities that might otherwise find use as vaccine centers [8]. Toyin Saraki, who heads the Wellbeing Foundation Africa, notes how Boko Haram has wiped out about 75% of the basic infrastructure required for immunization, including refrigeration, an essential in this hot and dry region, and personal health records and other monitoring systems, needed to ensure that children receive their vaccinations on schedule [9]. In addition to polio, measles, whooping cough (pertussis), bacterial meningitis, and even yellow fever have returned into the midst of the region's collapsed vaccine infrastructure [10]. Indeed, a 2019 study found that among children who reside within 10 km of an armed conflict area, their likelihood of receiving any vaccine declines by almost 50% [11]. As eloquently and simply stated by Eunice Damisa from the National Polio Emergency Centre located in Abuja, the capital of Nigeria, "Boko Haram does not accept the basic premises of medical science" [10].

Unsurprisingly in this setting, in 2017 a cholera outbreak has also occurred in Borno State in northeastern Nigeria, with almost 6,000 cases and a case fatality rate that was highest among those with advanced age [12]. Peaks in the epidemic coincide with heavy rains and floods, so potentially climate change also played a role. As in the Middle East, war, poverty, and climate change combine in a toxic mix.

In 2019, Muhammadu Buhari won a second term as Nigeria's president, but he now governs a nation severely weakened by Boko Haram and other internal conflicts. Despite its oil and energy reserves and an economy that exceeds those of at least three or four G20 nations, the all-consuming conflicts and the presence of the Islamic State in West Africa ensure that Nigeria will remain ground zero from NTDs in the near future.

From Bad to Worse: South Sudan, DR Congo, and Central African Republic

The region encompassing South Sudan, DR Congo, and Central African Republic now represents an area of conflict and disease as bad as any worldwide. Because of the lack of roads and infrastructure in these three nations, it may take us years to fully realize the horrors now being generated in this part of the world, but we are already getting glimpses of direct deaths attributable to war, collapsed health systems, and child malnutrition, with a simultaneous rise in NTDs.

The complicated politics and civil war between rivals in South Sudan have produced 400,000 deaths and not much hope for future stability, although there is now a signed cease-fire agreement between the South Sudan president and his major rival, a rebel leader, until the year 2022 [13]. In the meantime, an estimated four million people have been displaced, roughly divided in half between those internally displaced versus those fleeing to Uganda, Sudan, Ethiopia, and other neighboring countries in eastern Africa. Agriculture has also collapsed, so famine is widespread and brings with it heightened susceptibility to severe disease.

From this setting has arisen an awful NTD known as kala-azar. Also known as visceral leishmaniasis, the disease differs clinically from the cutaneous leishmaniasis cases found in Islamic State–occupied zones, even though the underlying social determinants promoting the rise of these two diseases are similar. In both Africa and the Middle East, collapsing infrastructures allow a proliferation of sand flies that inject parasites. Whereas the major *Leishmania* species found on the Arabian Peninsula, including *L. tropica* and *L. major*, cause mostly a disfiguring skin infection, in South Sudan, *L. donovani* causes serious systemic disease—a fatal illness that resembles a leukemia or lymphoma. After its inoculation by *Phlebotomus* sand flies, the parasite invades an important immune defense cell in our bodies called a macrophage. Although the macrophage is a type of cell equipped with destroyer enzymes and the ability to generate toxic chemicals that kill invading bacteria and parasites, miraculously the *Leishmania* parasite has evolved to survive inside the macrophage and therefore spread into human organs wherever large numbers of macrophages concentrate. This typically means the liver and spleen, which become enlarged and diseased, and the bone marrow. In these target organs, *Leishmania* parasites produce severe illness. I think of it as a type of parasite-induced blood cancer, except for the fact that we can treat it successfully with prolonged courses of antiparasitic drugs. However, in South Sudan, the drugs are often not available, so infected individuals ultimately succumb and die from this disease. *Leishmania* parasites also proliferate especially well in children and in both children and adults who are malnourished, because malnutrition interferes with normal immune functions. With the collapse of agriculture in South Sudan, malnutrition and famine are at all-time highs, and in tandem, widespread and deadly kala-azar infections.

Approximately 100,000 people died from a large and lethal epidemic of kala-azar that began in the 1980s and continued for two decades before ending with a comprehensive peace agreement in 2005 [14]. Starting in 2005, the agreement allowed the southern region of Sudan to break away and form a new state of South Sudan. However, in the early years after the peace agreement and up until formal independence was achieved in 2011, an already fragile health system collapsed, resulting in thousands of new kala-azar cases [14, 15]. In 2013, the newly independent South Sudan nation entered into yet another period of civil war. During the early 2000s, much of the treatment of kala-azar patients was provided by Médecins Sans Frontières (MSF) before the WHO helped to establish a network of treatment facilities that were run and managed by additional nongovernmental development organizations [15].

Kala-azar is especially lethal in patients with underlying malnutrition or those coinfected with HIV/AIDS. Widespread food insecurity in South Sudan, together with transmission of HIV/AIDS, the lack of easy access to diagnosis and treatment facilities, and extreme poverty linked to inadequate housing and outdoor exposure, further complicate the kala-azar epidemic in this part of the world [15–17]. It has also been noted that some factors unrelated to conflict are associated with an increased risk of kala-azar, including living in proximity to dogs, which serve as animal reservoirs of the disease, or sleeping next to or under acacia trees, where sand flies proliferate [15–17].

Along South Sudan's western border are two nations similarly beset by long-standing conflict and wars—the Central African Republic and DR Congo. Both nations, in addition to South Sudan, have endured epidemics of African sleeping sickness caused by a single-cell parasite in the blood and central nervous system, known as a trypanosome [18]. Instead of a sand fly, sleeping sickness is transmitted by a tsetse, which superficially resembles a common housefly but with the ability to produce a painful bite [18]. After the tsetse inoculates trypanosome parasites, they replicate in the bloodstream before entering the central nervous system to cause fatal sleeping sickness [18]. With some cessation of hostilities in both Central African Republic and DR Congo since 2016, it has been possible to conduct case detection and treatment programs or to enact tsetse control in order to reduce the prevalence of trypanosomiasis infection to just over a few thousand cases [19]. On that basis, the WHO, together with the Gates Foundation and other organizations, is working toward the possibility of eliminating trypanosomiasis from Africa in the coming years.

Ebola Virus Infection in DR Congo

Ebola virus infection made headline news in 2014 when it caused a sweeping and deadly epidemic across a part of western Africa, focused in the countries of Guinea, Liberia, and Sierra Leone. The health system collapses resulting from years of civil war and atrocities in these countries helped to ignite Ebola virus infection, which began in a small village in Guinea to spread into the capitals of Liberia and Sierra Leone. For the first time, Ebola became widespread in urban centers. The epidemic spread to at least seven countries, including Texas in the United States, with fears that it would spread widely in Nigeria, especially in the Boko Haram areas in the northeast. Ultimately, an international response ensued after the WHO declared a public health emergency of international concern (PHEIC) [20], prompting multiple nations to mobilize their militaries and healthcare expertise so as to provide the essential elements of a healthcare infrastructure. The WHO declares a PHEIC when a committee of experts decides an "extraordinary event" is under way, necessitating an increase in preventive measures at international border crossings, together with a request for a coordinated international response. This provision is an important component of IHR (2005). During the Ebola PHEIC in western Africa, a consortium of vaccine experts provided early proof-of-concept for the potential efficacy of an Ebola virus vaccine, which was licensed to Merck & Co. By the end of 2015 and early 2016, the implementation of an infrastructure for isolating and treating Ebola patients helped to end the epidemic, but not before almost 30,000 people became infected and over 11,000 people died [20].

The 2014–15 Ebola epidemic in Africa illustrates many of the Anthropocene forces that are now driving twenty-first-century epidemics on the Arabian Peninsula and in northern Nigeria. They include unchecked migrations of human populations into cities where crowding outstrips urban infrastructures, and the collapse of health systems following devastating civil and international conflicts. These forces went hand-in-hand with another important driver—deforestation. Daniel Bausch, now at Public Health England, reported how deforestation related to population expansion in southeastern Guinea brought humans into contact with fruit bats, which serve as a natural reservoir of the Ebola virus [21]. A small outbreak in that area eventually spread to Conakry, the capital city of Guinea, before it reached the urban centers of Liberia and Sierra Leone.

Bausch illustrates nicely the role of twenty-first-century Anthropocene forces in igniting this outbreak. Profound poverty and economic desperation forced human populations into forested areas in order to find wood for producing charcoal, as well as to engage in the practice of mining and mineral extractions. In addition, up to 60,000 refugees fleeing civil wars in Sierra Leone and Liberia, as well as the Ivory Coast, entered into some of these same forest areas [21]. These practices increase human exposure to bats, a major animal reservoir of human viruses, including the Ebola virus. In parallel, climatological changes that prolong the dry season further contribute to deforestation, while impoverished populations hoping to improve their economic opportunities and access to healthcare travel back and forth into urban cities and slums, even though the slums cannot accommodate the influx of migrants. As a result, these newly urbanized populations go without access to clean water or adequate food and healthcare. The new resource-poor cities serve as templates for the spread of infections among urbanized populations.

Although western Africa has so far suffered from the world's largest Ebola epidemic, Ebola virus is actually believed to have first emerged as a human virus infection in the Republic of Zaire during the 1970s. Zaire became a new sovereign nation and dictatorship in central Africa in 1971, when Mobutu Sese Seko seized power through a military coup and nationalized the former Belgian Congo. These events destabilized the Congolese economy, and as I noted previously, facilitated the return of devastating trypanosomiasis epidemics that resulted in the deaths of thousands. The destabilization of the region also helped to promote the emergence of Ebola virus infection, first detected and isolated in 1976.

Ultimately, the Mobutu regime collapsed, and Zaire became the DR Congo in 1997. However, the eastern areas of the nation became highly destabilized yet again following the genocide in Rwanda in 1994 on DR Congo's eastern border. Beginning in 1998, the Kivu area, bordering the countries of Rwanda and Uganda, entered a period of military conflicts that lasted for two decades. As elsewhere in the Middle East and Africa, the consequences of long-standing conflict were predictable, as disease, famine, and widespread sexual violence contributed to the deaths of many children under the age of five.

By August 2018, many of the elements responsible for the emergence of Ebola virus infection in western Africa were now in place in eastern DR Congo [22, 23]. As groups began to flee political violence, they again came into contact with Ebola-infected fruit bats in areas of deforestation, poverty, and health systems collapse. As of July 2019, there have been more than 2,600 Ebola virus cases and approximately 1,600 deaths, prompting the WHO to declare a second PHEIC. By the end of 2019, there were more than 3,300 confirmed infections and 2,200 lives lost.

I believe that the number of Ebola cases and death toll would be far higher were it not for the fact that the Ebola vaccine, which first underwent development and clinical testing during the

2014–15 epidemic in western Africa, became available to vaccinate populations in eastern DR Congo. However, the conflict and political instability in the region make it extremely challenging to set up vaccination centers, together with efforts to monitor both the efficacy and safety of the vaccine. Therefore, we face a situation with two opposing forces—vaccination programs versus conflict; waging peace with vaccines versus war.

From my perspective, it is almost miraculous that more than 200,000 individuals received the Ebola vaccine under nearly impossible conditions of war and distrust. I believe this is an extraordinary public health triumph in the making, namely, the development, testing, and deployment of Ebola vaccine through concerted efforts of the UN and US Department of Health and Human Services agencies, the Congolese Government, MSF, and the Wellcome Trust, to name just some. As details emerge, this will become one of the truly great stories in vaccine diplomacy. Equally impressive is the fact that cholera vaccine is also being deployed following an outbreak in Kivu [24], which might also have been predicted given the circumstances of war and health systems collapse we saw in Yemen. On the other hand, it is sobering to note that measles vaccinations have halted in many places across the DR Congo. According to some estimates, measles has so far affected close to 250,000 people and killed 5,000, about twice the number who died from Ebola virus infection. The story of vaccine diplomacy in DR Congo is notable for its extremes in terms of both failures and successes.

7:

The Northern Triangle and Collapse of Venezuela

Even in the absence of outright war and conflict, profound political and socioeconomic instability may also lead to disease. We are seeing this situation play out in the New World, especially in Latin America: Central America, Venezuela, and its neighbors. Our National School of Tropical Medicine at Baylor College of Medicine works globally, but because of their proximity to Texas, many of our activities focus on Mexico and central Latin America (sometimes also referred to as Mesoamerica). For example, we are developing a new therapeutic vaccine for Chagas disease, which might one day be used as a biologic treatment alongside existing antiparasitic drugs. Chagas disease transmission is widespread in Texas, Mexico, and central Latin America, and we are conducting vaccine development jointly with a consortium of Mexican scientific institutions [1]. Our faculty also conduct many studies in central Latin America. In fact, our associate dean, Dr. Maria Elena Bottazzi, who also co-leads vaccine development with me, is from Honduras, while Dr. Laila Woc-Colburn, our former head of clinical tropical medicine, is from Guatemala and Panama. In one way or another, most of our faculty seem to have some central Latin American connection. For example, Dr. Kristy Murray heads a CDC-funded unit to study viruses transmitted by mosquitoes (also known as "arboviruses," a portmanteau of "arthropod-borne virus").

Central America

Central America has a special allure and interest for me. While so much global health attention focuses on sub-Saharan Africa, I'm impressed both by Central America's natural beauty and its almost equal depth and breadth of poverty and pervasive poverty-related neglected diseases. What is especially striking is the close proximity of Central America. In about the time it takes to fly from Houston to the east or west coasts of the United States, I can be in a Central American capital. After a three- to four-hour flight, our research teams can investigate NTDs.

Whereas much of the Latin American region has made great strides in disease control, the Northern Triangle, comprising El Salvador, Guatemala, and Honduras, has stalled in its progress—and in some cases, diseases are rising or returning. For example, in the Northern Triangle nations, as well as in Nicaragua, the rates of dengue and other arbovirus infections are growing to the point where they are practically ubiquitous among the population, while the rates of other NTDs, including Chagas disease, leishmaniasis, and parasitic worm infections, have also increased [2]. One of the major reasons for this rise is a new level of political instability. Over the past decade, many of the routes used for the smuggling of drugs and other contraband that once went through the Caribbean and Florida now flow through the Northern Triangle. This shift has fueled the formation of brutal drug cartels that beget forced criminal gang recruitment of children and young adults, heightened gender-based violence, and extortions or kidnapping [2]. The escalating havoc has produced economic downturns and internal displacements of populations, as well as human caravans through Mexico.

In parallel, a prolonged drought throughout this region has created a so-called dry corridor cutting through the Northern Triangle, helping to collapse agriculture and promote food insecurity [2]. Along this strip, there has been a spike in acute and chronic kidney failure of unknown origin, especially among some of the agricultural workers. The causes of this Mesoamerican nephropathy are under investigation by Dr. Murray and her team, but she is looking into both infectious and environmental factors. The dry corridor has its origins in climate change and is a cause of food insecurity and downturns in agricultural economy, thereby further reinforcing the political instability and drug violence.

Exactly how political instability, violence, and climate change combine or interact to promote disease still requires intensive investigation, but it is troubling to see how Central America has become a hot zone for tropical infections—one that is very close to the southern border of the United States. So far, we have not seen a significant return of measles or other vaccine-preventable diseases, as in the Middle East or sub-Saharan Africa, but this aspect is something we must continue to monitor.

The Bolivarian Revolution

Beyond Mesoamerica, we are also working extensively in Panama and in tropical nations of South America, including Colombia, Ecuador, and Brazil. Sandwiched between these nations sits the Bolivarian Republic of Venezuela, now suffering one of the worst financial collapses in modern times for a nation not at war. The financial collapse and resulting instability that have led to widespread crime, corruption, malnutrition, and emerging disease are happening on a scale that may exceed some of the better-known historical socioeconomic crises, such as the breakup of the Soviet Union or even the Great Depression in the United States. These conditions have also led to the world's largest refugee crisis. By some estimates, more than three million refugees have left Venezuela, approximately 10% of the nation's population, with many emigrants spreading diseases acquired at home into neighboring Brazil and Colombia, ultimately destabilizing South America.

During the early 2000s, the country's then-president, Hugo Chávez, invoked the name of Latin America's nineteenth-century revolutionary leader, Simón Bolívar, to glorify his socialist regime and the conversion of many of Venezuela's private industries to a state-led economy. However, after his election in 1998 and formation of a new Venezuelan constitution the following year, the national economy began a long and inexorable decline, greatly exacerbated by declining oil prices. Because Venezuela's oil reserves and exports are central to its economy, the collapse of energy prices had a devastating effect. Political corruption and widespread breakdowns in national security then followed. By 2013, with the succession of Nicolás Maduro following the death of Chávez from cancer, up to 1.5 million people, or 6% of the nation's population, had left the country. Five years later, that number had doubled to 3 million, according to the United Nations High Commissioner for Refugees.

Maduro's 2013 election preceded a precipitous decline in public sector support for health and food security, resulting in chronic shortages in essential medicines and food. Hunger became widespread. Unemployment skyrocketed as Venezuela's economy began to implode and companies closed, forcing up to 90% of the population to live in poverty. In parallel, human rights violations became commonplace, including thousands of extrajudicial executions. Crime became rampant, and Venezuela now has one of the world's highest homicide rates.

By January 2019, when Maduro began a new term as president, his regime was widely condemned internationally. The Organization of American States designated it a dictatorship. In

2018, the OAS issued a detailed report on the Maduro authoritarian regime, while the International Criminal Court began investigating the Venezuelan government's crimes against humanity. In turn, Maduro cut off diplomatic ties with the nations that challenged his legitimacy or recognized his major opponent, the leader of the Venezuelan National Assembly, Juan Guaidó. US President Donald Trump both condemned Maduro's regime as a dictatorship and openly discussed the possibility of US military intervention against Venezuela while meeting with leaders of Latin American countries at the UN General Assembly.

Halt in Vaccination, Internal Displacement, and Exodus: Vaccine-Preventable Diseases

Among the first illnesses to return with the collapses of regimes in the Middle East and Africa are the vaccine-preventable diseases, and the same is true for Venezuela. In collaboration with Alberto Paniz-Mondolfi at the Venezuelan Institute of Biomedical Research in Barquisimeto, and other Venezuelan and international investigators, I reported on the resurgence of vaccine-preventable diseases [3]. Measles is the first to return because of its high reproduction number, in part owing to its hardiness, which allows it to linger in the atmosphere or on inert surfaces. As the Venezuelan economy began to unravel, so did its health systems and epidemiologic surveillance systems for monitoring vaccines. Beginning in 2010, there were increasing numbers of interruptions in national vaccine programs, together with piecemeal collapses in Venezuela's primary healthcare system [3]. Vaccinations began to decline precipitously. Beyond the lapses in vaccination programs, Venezuela's mining camps, many of which were illegal, became venues for internally displaced unvaccinated workers to cluster and ignite additional measles outbreaks. Internally displaced populations came into contact with Venezuela's indigenous populations, such as the Yanomami, or in some cases the indigenous workers themselves began working in the mines. Tragically, the Yanomami and other indigenous people were highly susceptible to measles, much as Native North Americans were when they first experienced contact with European explorers beginning hundreds of years ago.

The results were predictable. Even though measles was once eliminated from Venezuela, it reemerged in 2017. By the end of 2018, there were more than 8,000 new measles cases in Latin America, with two-thirds in Venezuela [3]. At least 80 people died from the disease. Based on examination of the specific genome of the measles virus, the overwhelming number of Latin America's cases either occurred in Venezuela or were a direct result of spread from refugees fleeing Venezuela to neighboring Brazil, Colombia, and other South American countries [3]. In Brazil, the two states bordering Venezuela—Amazonas and Roraima—were the most affected. Diphtheria, another serious and potentially deadly vaccine-preventable disease, has also emerged in the mining camps and is spreading into other areas of Venezuela and neighboring Brazil [3]. There are further concerns that polio could resurface in these areas, and now we're seeing the emergence of COVID-19.

In September 2018, the UN Human Rights Council proposed and adopted a resolution on Venezuela's humanitarian crisis, which included the urgency of addressing the rise in vaccine-preventable diseases and underlying malnutrition [3]. Around that time, I made a site visit to La Guajira, a peninsula on the northern coast of Colombia, located next to its eastern border with Venezuela. La Guajira is a highly arid and impoverished region that is host to several indigenous peoples, including the Wayuu, a group famous for their historic resistance to colonization and conquest by the Spanish. Today, many of the Wayuu live in profound poverty and suffer from high rates of malnutrition, owing in part to the difficulties of growing produce in such a dry environment. Texas Children's Hospital and Baylor College of Medicine are working in La Guajira to support nutritional programs among the Wayuu, and because measles is especially lethal in malnourished children, I became concerned about the potential vulnerability of Wayuu children to measles and other vaccine-preventable diseases spreading from Venezuela to Colombia.

The sharp rise in measles and diphtheria, together with underlying malnutrition, produced a humanitarian catastrophe for the people of Venezuela and neighboring countries. There is an urgent need to pursue emergency relief operations for catch-up vaccine programs, along with reestablishment of an adequate supply chain for vaccines and implementation of epidemiologic surveillance systems in the region [3]. The WHO has yet to declare a PHEIC in Venezuela, and there is a need to change Venezuela's eligibility requirements for receiving free vaccines [4]. To receive free vaccines, a nation has to fall below specific economic indicators, and Venezuela has not yet met those criteria. In turn, the Maduro government has resisted international calls for humanitarian interventions.

Halt in Vector Control and Disease Introduction: The Neglected Tropical Diseases

The Venezuelan collapse has also brought back tropical infections, especially those transmitted by mosquitoes and other arthropods. The return of malaria has been dramatic. At the start of the twenty-first century, almost all of the Latin American nations committed to the UN Millennium Development Goals for malaria by embarking on extensive control efforts. Between 2000 and 2015, the region saw symptomatic malaria cases and deaths decline by more than 60% [5]. However, Professors Maria Grillet and Belkisyolé Alarcón-de Noya and a group of Venezuelan colleagues and collaborators and I reported how Venezuela saw a 359% increase over that same period, and then another 71% increase from 2016 to 2017 [5]. In addition, Brazil saw a sharp uptick in imported cases from Venezuela.

The true increase in malaria cases and deaths in Venezuela is unknown, because a high percentage of malaria transmission occurs in remote or illegal mining camps. Indeed, in terms of malaria, mining is now a high-risk profession. Clearing forests creates conditions that are favorable for breeding the *Anopheles* mosquitoes that transmit the disease, while the miners themselves become highly vulnerable to mosquito bites as a result of sleeping outdoors without bed nets, and because they often emigrate from areas of Venezuela where malaria transmission does not exist [5]. Such internally displaced populations are immunologically naive, meaning that they acquire malaria for the first time in the mining camp [5]. Often, first infections with malaria are severe or even fatal. Further hampering malaria control and containment efforts is the disappearance of government-led interventions to conduct malaria surveillance and report on its emergence, together with government failure to procure antimalarial drugs, as well as bed nets, insecticides, or diagnostic kits—in other words, all of the essential tools needed to fight malaria [5].

Several other parasitic diseases are also returning to Venezuela. Just as leishmaniasis returned to the Middle East and eastern African conflict zones, it is also rising in the wake of Venezuela's political and socioeconomic collapses. Both cutaneous and visceral leishmaniasis occur, although they are caused by New World species of parasites (and transmitted by New World sand flies) that differ from those I described previously. The mechanisms by which leishmaniasis is reemerging are also not too different from what I described earlier for Ebola virus infection in Guinea in 2014. Deforestation and urbanization brought populations into contact with animal vectors of disease—bats in the case of Guinea and DR Congo, and sand fly vectors in the case of Venezuela. Bringing together humans and sand flies created opportunities for peri-urban transmission [5].

Transmission of still another parasitic infection, Chagas disease, occurs uniquely in the New World, predominately in the poorest regions of central Latin America. Kissing bug vectors, also known as triatomines, transmit a trypanosome protozoan that superficially resembles the African trypanosome responsible for sleeping sickness. However, unlike its African counterpart, the American trypanosome—*Trypanosoma cruzi*—has the ability to invade the human heart and cause a debilitating and sometimes fatal disease known as Chagasic cardiomyopathy. Six- to seven million people in the Americas live with Chagas disease, including more than one million with Chagasic heart disease. Most Chagas disease patients subsist in extreme poverty and without access to antiparasitic drugs. Venezuela had previously made steady progress in controlling Chagas disease through both the collateral effects of insecticidal spraying for malaria, together with improvements in housing. Housing improvements are essential, since the kissing bug has the ability to live in the cracks and crevices of low-income dwellings, especially those in tropical regions that have a thatched roof. Nevertheless, by the late 1990s, government-led surveillance and control efforts began to fall apart, and by 2012, they halted altogether [5]. In some Venezuelan villages it was estimated that almost one-quarter of the population was infected with *T. cruzi* at the time that control programs began to lapse. Similar to leishmaniasis, peri-urban transmission of Chagas disease is occurring in deforested areas where desperately impoverished populations seek access to food and safety. Yet another issue is the problem of orally transmitted Chagas disease [5, 6]. More than a dozen outbreaks have been linked to infected kissing bugs contaminating containers of fruit juice, presumably when they are crushed together with the fruit. The fact that triatomines now have regular access to Venezuela's food supply reinforces my impressions about the profound decline in the economy and security.

Venezuela also now represents one of the last nations in the Americas where schistosomiasis is prevalent. Schistosomiasis caused by *Schistosoma mansoni* is a serious cause of liver and intestinal disease that results when these parasitic worms deposit their spine-shaped eggs in those organs to produce inflammation and damage. Freshwater snails transmit the disease, so individuals acquire schistosomiasis through wading, bathing, fishing, or washing clothes in infested rivers, streams, and lakes. *Schistosoma mansoni* infection was believed to be introduced into the Americas between the sixteenth and nineteenth centuries, coinciding with the Atlantic slave trade. Today, most of Venezuela's schistosomiasis cases occur on the northern coast, but surveillance and control activities through mass drug administration with praziquantel have mostly halted, so it is difficult to know for sure.

Similar to the other disease-control programs, efforts to halt the spread of arboviruses—viruses transmitted by mosquitoes and other arthropods—have also lapsed. Of particular concern are the those transmitted by the *Aedes aegypti* mosquito. This mosquito species is pervasive in tropical and subtropical areas of the Americas and is particularly well adapted to urban slums, where environmental degradation is extensive. The mosquito lives in old tires and other discarded containers that collect rainwater.

Even well-maintained areas in the American tropics struggle to combat the major arbovirus infections, especially dengue. Now the poor urban areas of Venezuela face a particularly dire situation. One unique aspect of the Venezuelan crisis is the interruption of city water supplies, forcing many families to collect or store water either in their homes or outside [5]. Containers of stored water provide excellent homes for *Aedes aegypti*. As a result, dengue has now spread widely across urbanized Venezuela, where it is a serious cause of morbidity and even mortality. Then, beginning in 2013, two new arboviruses entered the picture. First, chikungunya and, later, Zika virus emerged in the Western Hemisphere to cause an additional serious public health threat in Venezuela. According to some estimates in 2014, more than two million people became infected with chikungunya, many of which cases were more severe than generally reported, and some were fatal [5]. Two years later, over the first eight weeks of 2016, Zika virus infection swept through Venezuela, infecting a high percentage of pregnant women and causing a significant number of cases of congenital birth defects, as well Guillain-Barré syndrome, a serious neurological complication that can result in paralysis [5].

The scientific basis for the sharp rise in chikungunya and Zika is still under investigation. These two viruses may have entered the Americas through different routes. There is some evidence that Zika virus underwent genetic changes as it traveled east across the Pacific Ocean before emerging in South America in 2013 or 2014, but it is not clear that mutations also explain the rise of chikungunya, which first emerged on the island of Saint Martin in the Caribbean around the same time. In the case of chikungunya, it may have traveled west across the Atlantic Ocean from Africa. Once introduced, both of these virus infections spread very quickly across the Latin American and Caribbean region, and even entered southern Florida and southern Texas in the continental United States. Also facilitating the spread of Zika and chikungunya was the fact that the Latin American and Caribbean population had never encountered these viruses before and therefore were immunologically naive, meaning they had no preexisting immunity.

The rise of arboviruses in Venezuela also threatens to destabilize the Caribbean and Central and South America. And the arbovirus diseases arising from Venezuela's economic collapse have even assumed a global aspect. For instance, dengue has emerged on the island of Madeira, off the coast of Portugal, but it may have originated in Venezuela [5, 6]. The introduction of dengue into Europe may be just the start of health consequences from the Venezuelan diaspora. We should anticipate how arboviruses might eventually enter other European countries, as well as major US cities such as Miami, Tampa, and Houston—which host large numbers of Venezuelan immigrants and asylum seekers.

Latin American Vaccine Diplomacy

Imagine a poor laborer no longer able to find work in his native village who is forced to work far away in illegal mines in order to provide for his family. He is continuously exposed to mosquito bites and acquires malaria or dengue, or he contracts leishmaniasis from sand fly bites, or Chagas disease from food contaminated by kissing bugs. He also faces mercury exposure and poisoning because this metal is used to extract gold, and then he is at risk for measles exposure. When he returns home or emigrates to Brazil or Colombia, he brings these illnesses with him. Along his travel routes, he exposes indigenous populations, or the indigenous populations themselves work in these conditions. This scenario now occurs on a daily basis in Venezuela.

Possibly the group most affected by the Venezuelan collapse is the Yanomami indigenous people. There are an estimated 40,000 people living in and around the Venezuela-Brazilian border, many in extreme isolation. Internally displaced workers or refugees fleeing Venezuela and traveling through Amazonas State in Brazil will contact this population and transmit disease. Moreover, Yanomami indigenous populations are now working in some of the illegal mining camps and then returning to their villages to introduce diseases there. There are reports of high death rates from measles, malaria, and malnutrition among the Yanomami. The first case of COVID-19 was recently reported in a Yanomami boy [7].

A humanitarian tragedy is under way because of the rise of malnutrition and infectious and tropical diseases in Venezuela. Is there a role for international vaccine diplomacy? As Venezuela becomes a failed nation-state, and now one in which its leaders sometimes refuse humanitarian assistance, this will be no easy task. The current situation offers opportunities for citizen scientists and informal networks of healthcare providers to fill the current gaps in epidemiologic surveillance, which in turn provides essential disease prevalence and incidence data, but this is a far from optimal situation. Leaders from the Organization of American States, Pan American Health Organization, and WHO already work with the Venezuelan leadership to restore health systems, but progress on this front has been incremental at best.

As a scientist and science administrator working in Houston, Texas, one of the major gateway hubs for asylum seekers, I have now met with several Venezuelan scientists who either have been forced out by government authorities or whose laboratories have collapsed because of lack of funds and inabilities to maintain essential equipment or purchase supplies. I know of at least one scientist whose laboratory was ransacked and destroyed. Most of the scientists now fleeing Venezuela are still eager to continue working, but it is not easy, especially for midcareer and senior scientists, to start over and find a position here.

A preferred situation would be for US scientists to work with Venezuelan scientists to build capacity in Venezuela [8]. This could include joint development of vaccines for Venezuela's emerging NTDs, including the ones we are working on in Texas—schistosomiasis, leishmaniasis, and Chagas disease [8, 9]. With the US government recently threatening military intervention in order to potentially effect regime change in Venezuela, conducting vaccine diplomacy may simply have to wait until President Maduro either leaves office or reinvents his policies for international cooperation. However, I still believe it is possible to begin implementing vaccine diplomacy now. Among the steps to consider would be working with the US State Department and US Agency for International Development to support joint collaborations between US and Venezuelan scientists. This could be done as a public-private partnership, or potentially with a nonprofit foundation. Initially, those collaborations might begin in US research institutions and universities to provide a stable working environment for Venezuelan scientists, but support would eventually transition to Venezuelan institutions. This approach still does not address downstream vaccine development, but for this purpose, it might occur through a three-way partnership to include potential Latin American nations with this capacity, such as Argentina, Brazil, Cuba, or Mexico. The collapse of the Venezuelan nation is unprecedented in modern times. We will need extraordinary measures to restore disease control and scientific infrastructure.

8: Sorting It Out

Attributable Risks

A common denominator in the three case studies of the Middle East, Africa, and Venezuela, respectively, is how multiple social determinants—war, political instability, urbanization and deforestation, human migrations, and shifting poverty, as well as other Anthropocene forces, including climate change, combine to produce new areas of disease hot spots. Is it possible to determine whether any one particular factor dominates as the major driver of disease emergence? In the field of epidemiology, the concept of "attributable risk" refers to the portion of disease that we can attribute to a specific exposure. Attributable risk can also determine the portion of a favorable outcome of a disease treatment that can be credited to a specific intervention. Can we also assign attributable risk to the problem of complex disease emergence?

Disease in the New World

In Venezuela and its neighbors Brazil and Colombia, socioeconomic collapse halted national disease vector control and vaccination campaigns, while greatly limiting healthcare access. Internal population displacement, resulting from attempts to flee economic oppression, also became widespread and forced laborers into illegal mining camps, where they were especially vulnerable. Tropical infections and vaccine-preventable diseases returned. In parallel, extreme drought from climate change collapsed agriculture. It promoted more internal displacement, as well as unchecked urbanization, which outstrips cities' infrastructures and sewage management and leads to upticks in diarrheal disease and possibly urban helminth infections. Indigenous people such as the Yanomami are highly susceptible to common viral and bacterial pathogens because of their history of extreme isolation. Now through mining or through internal displacement of populations carrying disease, the Yanomami have experienced unprecedented contact, furthering disease exposure risks. Therefore, in this complex system, political collapse and climate change are mutually reinforcing to promote extreme poverty, unchecked urbanization, and widespread disease. Holding any single factor responsible for the rise of disease in Venezuela is highly challenging.

On a smaller scale, some of the same forces are playing out in more advanced economies, such as the state of Texas [1]. Despite the fact that Texas has a population size and economy approximately the same as Australia or Canada, it also has pockets of intense poverty, especially along its border with Mexico and in some urban locations. Currently, Austin, Dallas, Houston, and San Antonio rank among the 10 largest metro-area populations in the United States. Many people are surprised to learn that more than 85% of the population of Texas now lives in urban areas, a much higher percentage than other southern states and more akin to states in the Northeast. However, additional estimates indicate highly uneven economic growth in these cities. The urbanized poverty juxtaposed with tremendous wealth in the four major Texas cities places the state near the top in disparities between wealthy and impoverished populations, as reflected in an economic value known as the Gini coefficient—a metric used by economists for this purpose. Texas represents one of America's most extreme examples of diseases of the poor amid wealth [1]. In the "colonias" (unincorporated communities on the Mexican border, often lacking adequate sewage management) [1] or in the poor neighborhoods of Texas cities, it is common to find perfect conditions for breeding urban-dwelling *Aedes aegypti* mosquitoes or other disease vectors. Fueling the rise of arboviruses and vector-borne disease is a growing threat from climate change, with a possible doubling projected of days with temperatures exceeding 95°F [1]. Still another factor is the observation that ports along the Gulf Coast of Texas are expanding in order to accommodate a recent doubling of sea traffic through the enlarged Panama Canal. This additional circulation of cargo and humans by sea will also introduce new disease pathogens. Our scientists at the National School of Tropical Medicine at Baylor College of Medicine have noted a significant level of tropical infectious diseases in Texas, including most of the arbovirus infections, typhus, Chagas disease, multiple tick-borne infections (including relapsing fever), and some helminth infections. Tuberculosis is a major problem, especially in those with underlying diabetes. COVID-19 disproportionately affects poor neighborhoods, where implementing social distancing is difficult and where obesity, diabetes, and hypertension are widespread among African American and Hispanic communities. Texas also now has the largest numbers of children not receiving vaccinations in the United States. In this respect, the state represents the epicenter of anti-science in America. Demonstrations defying social distancing and claiming that COVID-19 is "fake news" have recently been staged in front of the state capitol in Austin. Not surprisingly, in 2020 Texas emerged as one of the states most affected by COVID-19.

In both Venezuela and Texas, a perfect storm of forces—extreme poverty, upticks in human migrations and urbanization, climate change, and anti-science—likely work together to promote disease emergence. We do not have an approach to sorting out attributable risk for any single factor.

Disease in the Old World

A similar confluence of Anthropocene forces is playing out in the Old World. In southern Europe, vector-borne diseases have emerged or reemerged [2]. By some accounts, malaria is back in Greece and Italy decades after its eradication through vector control and economic development. Leishmaniasis transmitted by sand flies is another vector-borne parasitic infection. And multiple arbovirus infections have now either emerged or reemerged, including West Nile virus infection, dengue, chikungunya, Toscano virus infection, and Crimean-Congo hemorrhagic fever, as have Lyme disease and other tick-borne infections. For the first time, schistosomiasis is in Corsica. Once again, the situation is complex. Climate change and warming temperatures are significant factors in southern Europe, and countries such as France, Italy, and Spain have recorded their highest temperatures ever [2, 3]. Also, climate change is not occurring in isolation. Greece and several other southern European economies have undergone sharp economic downturns and recessions. Then there are the human migrations from the Middle East and North African regions traveling across the Mediterranean, which potentially could introduce disease. Like Texas, some southern European nations, such as Italy, are now epicenters of populism and anti-science movements. As a result, measles is now a major infectious disease in the region. COVID-19 has devastated the health and economy of Italy and Spain.

Similarly, the unprecedented heat and drought of the Middle East that has been linked to global warming is driving families into stressful and crowded cities. However, urbanization also parallels the collapse of public health across the conflict zones of the Arabian Peninsula. As a result, the "Aleppo evil" transmitted by sand flies is now widespread in urban areas in the war-torn regions, and this disease, as well as cholera and schistosomiasis, has become widespread in Yemen. Adding to this complex pattern are unprecedented levels of human migration. Refugees fleeing the conflict zones are spreading emerging infections to other Middle Eastern nations. Simultaneously, the annual Hajj and Umrah bring millions of people from across the Muslim world (and their illnesses) into Saudi Arabia and from there into other areas of the Arabian Peninsula. This latter factor likely explains the emergence of dengue in the region. Vaccine-preventable diseases are also returning to the Middle East conflict zones where vaccination programs have halted. For a while, Iran was a top-10 country in terms of the number of cases of COVID-19. Once again, identifying individual attributable risks becomes a complex and daunting challenge.

The "un-wars" of sub-Saharan Africa have promoted the emergence or urbanization of Ebola, leishmaniasis, cholera, and vaccine-preventable diseases. However, deforestation and climate change synergize with war and political instability as well. An overlay of extreme poverty also fuels the rise of disease, with projections indicating that by 2050, 40% of the world's poor will live in Nigeria and DR Congo, perhaps in just two megacities. Although anti-science is not yet a dominant force in Africa, we do not know if the anti-vaccine movement might one day travel from America or Europe to the most desperate areas of the African continent.

Sorting It Out

The social and physical determinants of the Anthropocene are appearing in batches and likely combining to drive up disease in multiple global hot spots. Shown in table 1 is a partial listing

Table 1. Diseases of the Anthropocene and Their Drivers

Region	Major diseases	Major drivers
New World		
Venezuela, Brazil, and Colombia	Measles Malaria and tuberculosis Arbovirus infections Leishmaniasis and Chagas disease Schistosomiasis COVID-19	Socioeconomic collapse Political instability Poverty Climate change Urbanization Human migrations
Texas	Arbovirus infections Measles Typhus Chagas disease and leishmaniasis Tick-borne diseases Helminth infections Tuberculosis COVID-19	Poverty and blue marble health Urbanization Climate change Human migrations and cargo movements Anti-science
Old World		
Southern Europe	Measles Malaria Arbovirus infections Leishmaniasis and Chagas disease Schistosomiasis COVID-19	Climate change Poverty Human migrations Anti-science Urbanization
Middle East and North Africa	Leishmaniasis Cholera Arbovirus infections Schistosomiasis Tuberculosis Zoonotic diseases COVID-19	War and conflict Human migrations Hajj and Umrah Poverty Climate change Urbanization
Sub-Saharan Africa	Ebola Cholera Arbovirus infections Malaria and tuberculosis	Poverty Urbanization War and conflict Human migrations

Note: The specific diseases in the second column are not correspondingly linked to the drivers listed in the third column.

of some of the major at-risk areas of the world where diseases are now emerging, together with a potential ranking of their major Anthropocene drivers.

It is becoming clear that the Anthropocene factors promoting the rise of disease in twentieth-first-century hot zones interconnect, but so far, we do not have straightforward approaches for mechanisms to untangle the individual drivers. Solving something as complex as the rise of vaccine-preventable and tropical infections will require substantive discussions between different academic disciplines or between government agencies. Pick any one of the hot spot areas highlighted above, and it soon becomes obvious that addressing problems in the area of disease will require microbiologists and virologists to collaborate not only with each other but also with political scientists, economists, sociologists and poverty experts, earth scientists, and urban planners—to name just a few specialists. Unfortunately, most of our universities and governments tend to work in silos, so there are few if any incentives for this kind of cross-dialogue and collaboration. Increasingly such interactions will become essential if we are serious about predicting or solving the complex patterns of twenty-first-century disease emergence or reemergence and therefore implementing vaccine diplomacy.

9: Global Health Security and the Rise in Anti-science

We face a daunting task of curing and preventing the new Anthropocene diseases arising through contemporary social and physical determinants. These same diseases are now affecting global security. This is an idea that is still not widely accepted, but the evidence base is accumulating. Among the most obvious examples are the disruption and destabilizing effects of highly lethal infectious disease outbreaks such as Ebola virus infection in western Africa in 2014, and subsequently in Dallas, Texas, that same year [1]. Even more notable, over the course of ten months COVID-19 destabilized Asia, Europe, and North and South America, and crashed a prosperous global economy. A key point is that lethal infections and pandemic threats can both arise from political instability and reinforce it. They are major social disrupters.

A Global Health Security Agenda

Throughout the last half of the twentieth century and into this new millennium, global policymakers have recognized the links between health and global security. Security was front and center on the agenda of UN diplomats when they met to form the

World Health Organization at the end of World War II. WHO's founding constitution specifically states, "The health of all peoples is fundamental to the attainment of peace and security and is dependent upon the fullest co-operation of individuals and States" [2]. The diplomatic achievement of creating the WHO has certainly paid off in terms of public health gains. It produced a string of important global health victories, especially the eradication of smallpox and the launch of the Expanded Program on Immunization during the 1970s, ultimately preventing tens of millions of deaths from deadly infectious diseases. However, exactly how those gains translated into security was somewhat elusive.

Those links solidified in the early 2000s, following the SARS epidemics in China and Canada [1]. The 2003 epidemic in China disrupted airline, hotel, and other businesses, and even threatened to weaken the entire Asian economy. It also almost halted the economic development of Toronto when it caused hundreds of cases and more than 40 deaths there. Ultimately, SARS destabilized China's leadership, especially when it was slow to share information publicly about the disease's beginnings, spread, and public health impact.

The absence of transparency or the appearance of a potential cover-up of a serious and potentially life-threatening SARS epidemic prompted in 2005 a broadening of International Health Regulations through IHR (2005). Implementation of the revised and updated regulations mandated unprecedented levels of government transparency through the release of real-time and pertinent information about epidemics by governments, particularly for specially designated PHEICs [3]. IHR (2005) comprises an agreement between all WHO member states, which are now required to cooperate in a program of global health security, together with eight core units, ranging from disease surveillance to strengthened diagnostic laboratory testing capacity to risk management [3]. Another important element includes the implementation of preventive actions at international borders, seaports, and airports in order to limit disease spread [3].

Today, the WHO loosely defines global health security as "activities required . . . to minimize the danger and impact of acute public health events that endanger people's health across geographical regions and international boundaries" [4]. Dr. David Heymann—a celebrated medical epidemiologist, former assistant director-general of health security for WHO, and currently a health policy expert at London's Chatham House—points out how IHR (2005) was "put to the test in 2007" after the government of Indonesia halted information sharing about a strain of bird influenza [1]. In part, this situation occurred because Indonesia was denied access to a vaccine derived from the virus strain it had provided to an influenza international surveillance network [1]. Accordingly, a revised pandemic influenza preparedness framework was created [1], and this in turn was tested again in 2009 with the emergence of H1N1 pandemic influenza.

Then in 2014 during the western African Ebola epidemic, the WHO and CDC created a new Global Health Security Agenda (GHSA), in collaboration with the UN Food and Agriculture Organization, the World Organisation for Animal Health, Interpol, other organizations, and 29 initial partner nations [5, 6]. It has since grown to include over 60 nations, with a focus on strengthening national capacities to detect and respond to human and animal health threats [6, 7]. Another element includes external assessments and peer-to-peer review to measure a nation's capacity for responding to infectious disease threats according to specific targets. The intent is to build a health security ecosystem based on setting national priorities, adequate mobilization of funds and other resources, and measurement of progress. In 2017, the "Kampala Declaration" in Uganda extended the remit of the GHSA until 2024, continuing with the major goals of detecting, responding, and ultimately preventing disease outbreaks using a multisectoral approach.

Now IHR (2005) and the GHSA are being further tested by the COVID-19 pandemic. Following declaration of a PHEIC at the end of January 2020, the leaders of China and the United States exchanged accusations of a lack of transparency and of cover-ups regarding the origins of the SARS CoV2. The Trump administration has further accused the WHO of favoring China and of complicity in obscuring early evidence for human-to-human transmission of the SARS CoV2, and it has enacted bans on travel from China, while halting immigration. The US government threatens to cut off funding for the WHO, just as COVID-19 has expanded into the Southern Hemisphere, as well as the major nations of Latin America, Africa, and Southeast Asia.

What about the Vaccines?

Both IHR (2005) and the Global Health Security Agenda focus significantly on disease surveillance, detection, and response, but vaccines and vaccine development are not necessarily front and center in these programs. However, in the past decade we have seen how rapid deployment of vaccines can make an important difference in terms of averting otherwise catastrophic outcomes of disease. Through vaccine diplomacy in the middle of complex public health emergencies, vaccines are becoming potent disease-fighting agents. Two very exciting uses of vaccine diplomacy include access to cholera and Ebola vaccine, respectively.

Cholera. Efforts to halt the spread of cholera during emergencies may now increasingly rely on vaccines and vaccine diplomacy. In 2010, we highlighted our concern that only about 400,000 doses of cholera vaccine were available globally. We then called on the US government to stockpile sufficient cholera vaccine for potential humanitarian disasters [8]. Cholera was then emerging in the aftermath of the Haiti earthquake. Although the US government did not take us up on the suggestion, an International Coordinating Group, composed of representatives from the WHO, UNICEF, MSF, and the International Federation of Red Cross and Red Crescent Societies led efforts to manage a Geneva-based global stockpile of cholera vaccine [9]. Gavi, the Vaccine Alliance finances the stockpile, working in close partnership with the Global Task Force for Cholera Control and distributes millions of doses in multiple nations annually. Then in 2018, the stockpile faced its biggest test yet when facing the world's largest cholera epidemic, which had arisen out of the Yemen conflict [10]. The outbreak began in 2017 and resulted in more than one million cases. Although some questioned why it took a year to mobilize the stockpile, ultimately a consortium of WHO, UNICEF, Gavi, and World Bank partners successfully deployed cholera vaccine at the start of Yemen's 2018 rainy season, after the Global Task Force for Cholera Control requested more than four million doses [11]. Such efforts are part of a larger attempt to end cholera transmission by 2030.

Ebola. An equally dramatic effort is under way to deploy the Ebola virus vaccine in a complex emergency. Toward the end of the 2014 western African Ebola virus infection epidemic, in which more than 11,000 people lost their lives, the Obama administration worked through its lead agency, the Biomedical Advanced Research and Development Authority within the US Department of Health and Human Services, to support and accelerate the development and stockpiling of new Ebola virus interventions. Among the most promising was a single-dose live virus vaccine known as rVSV-ZEBOV-GP, first developed by the Public Health Agency of Canada and then licensed to Merck & Co. The Biomedical Advanced Research and Development Authority provided substantial funding after the vaccine showed potential efficacy during the epidemic in Guinea, when it was used in what is referred to as a ring-vaccination protocol [12]. First pioneered during the smallpox eradication campaign, the process of ring-vaccination works by halting virus transmission by vaccinating contacts of a patient with the disease, and then vaccinating the contacts of the contacts [12, 13].

As the Ebola epidemic began to take off in the DR Congo in 2019, WHO's Strategic Advisory Group of Experts approved ring vaccination [14]. The goal was to avoid a catastrophic loss of lives, as had occurred in western Africa five years earlier. Because the area is beset by conflict, distrust, and political unrest, vaccinating Ebola contacts in the Kivu area of DR Congo is a daunting task. Tragically, a Cameroonian physician fighting the epidemic was murdered. Yet an April 2019 report from the WHO revealed that almost 100,000 people were successfully vaccinated in almost 700 rings [14]. Remarkably, the vaccine was 88.1 to 97.5% protective, depending on how the efficacy was measured [14]. From my viewpoint, the rapid deployment of the vaccine, together with extraordinary coordination between the US and Canadian governments, the WHO, a pharmaceutical company, Gavi support, and the health workers in DR Congo and the Congolese government, averted a catastrophe that might have approximated or exceeded the 2014 Ebola epidemic.

Beyond cholera and Ebola, in 2013 the WHO launched a new framework for vaccination in times of humanitarian emergencies, with a goal of assisting countries and organizations through the logistical and ethical hurdles and maximizing the numbers of lives saved. Based on a 2016 series of consultations, the framework added processes for decision making, implementation, and access to country case studies, and for the establishment of procurement mechanisms for affordable vaccines during humanitarian emergencies [15]. The examples of cholera and Ebola virus infection represent high-level international cooperation and modern-day vaccine diplomacy. Thousands of lives were saved in Yemen and DR Congo, respectively. The fact that vaccine diplomacy was implemented in some of the world's worst war-torn conflict zones is a testament to its power. Now, the WHO is partnering with nations to establish a framework for developing and distributing COVID-19 vaccines.

The Rise in Anti-science

Since 2015, a new threat to global health security and vaccine diplomacy has emerged, but one having little to do with war, conflict, climate change, or urbanization. An anti-vaccine misinformation movement and campaign that began as a fringe group in the early 2000s has gained sufficient critical mass to affect public health. By 2019, measles had returned to Europe with more than 100,000 cases, and for the first time in two decades, it had returned to the United States. In a measles epidemic in New York City, approximately 50 people were hospitalized, including 18 in an intensive care unit. Beyond measles, thousands of American teenagers were denied access to their human papillomavirus (HPV) vaccine and therefore cancer prevention, while thousands more Americans actually died because they chose not to vaccinate themselves or their children against influenza—despite recommendations from the CDC. The anti-vaccine movement has also started to globalize. At the end of 2019, several news outlets reported that a devastating measles epidemic on the Pacific island of Samoa was fueled by leaders of the anti-vaccine movement in America. It won't stop there.

I began a personal initiative to counteract the misinformation of the anti-vaccine movement in the United States, but for reasons having little to do with my role as US science envoy. I am not only a vaccine scientist and pediatrician, but also the parent of four adult children, including Rachel, now a 27-year-old with autism and intellectual disability. This information is relevant because a central tenet of the anti-vaccine movement is that vaccines cause autism, even though there is massive evidence refuting any link, or even plausibility, given what we have learned about the genetics, natural history, and developmental pathways of autism. Starting in 2016, I began writing and speaking about the dangers of the anti-vaccine movement because it was producing steep declines in the numbers of kids vaccinated. My actions prompted a massive backlash from the leaders of the anti-vaccine movement, and a steady stream of attacks directed against me in books, on social media, and even by stalking at meetings and other venues. Their aggression and tactics taught me how the anti-vaccine movement is well funded and well organized and has now become a threat every bit as potent as any of the Anthropocene forces described so far. It may be representative of additional anti-science movements to follow. However, we have an opportunity to dismantle these activities, possibly in time to prevent them from spreading further to Africa, Asia, and Latin America.

Measles and the Anti-vaccine Movement

After the global eradication of smallpox in the late 1970s, measles became one of the next big targets. During the 1970s and 1980s measles killed over two million children annually, yet through vaccination campaigns, initially the WHO's Expanded Program on Immunization, and later through the activities of Gavi, that number by the 2010s was brought down to just around 100,000. Both the global health community and I consider this a tremendous public health triumph.

However, in 2018 measles returned to Europe and to the United States in 2019. In Europe, estimates from the WHO indicated that more than 80,000 people contracted measles, resulting in large numbers of hospitalizations and at least 70 deaths [16]. The number of measles cases represented a 15-fold increase from 2016 (around 5,000), the year when Europe reported its fewest ever measles patients, but then the numbers started to climb to more than 20,000 cases in 2017. During the first half of 2019, Europe experienced an estimated 90,000 measles cases. Across the Atlantic in the United States, the CDC reported more than 1,000 measles cases, our highest measles numbers since 2000, when measles was eliminated [17].

In the cases of Europe and the United States, climate change and the social determinants of poverty and urbanization did not reintroduce measles. Even the wars and conflicts in the Middle East, which promoted refugee movements across the Mediterranean, accounted for only a tiny percentage of Europe's measles cases. Instead, the rise of measles was the consequence of declines in childhood vaccine coverage. In Europe, measles reemerged after vaccine coverage fell in several nations, especially in countries such as France, Italy, and Greece in southern Europe and then in Romania and Ukraine in the east. In the case of the United States, vaccine coverage declined sharply in my state of Texas [18], as well as in 17 other states that allowed vaccine exemptions for reasons of personal and philosophical belief [19].

The drop in vaccine coverage in the United States and Europe occurred mostly in the 2010s as a result of a growing misinformation campaign and selected political activities commonly referred to as the anti-vaccine movement [20]. In my previous book *Vaccines Did Not Cause Rachel's Autism: My Journey as a Vaccine Scientist, Pediatrician, and Autism Dad*, I detailed how the antivaccine movement started to take off following the 1998 publication of a paper alleging that the vaccine for measles, mumps, and rubella might cause autism [21]. It fell to Brian Deer, a British investigative journalist, to reveal how the paper was fraudulent—subsequently prompting its retraction—but the publication nonetheless disrupted confidence in the measles and other vaccines. By the 2010s, the anti-vaccine movement was in full swing.

Today the anti-vaccine movement, along with climate change denial and fear of genetically modified organisms (GMOs), ranks as one of the world's largest and possibly most dangerous anti-science activities. All three anti-science engagements have the potential to adversely affect public health, but the anti-vaccine movement is producing the most direct and immediate impact, in the form of a return of a serious infectious disease that is now causing hospitalizations and injury in the United States and Europe.

I identify three major components of this modern anti-vaccine movement [21]. First, it has developed from its beginnings in the early 2000s, as a fringe element, into its own media empire. I sometimes equate the anti-vax media group with something resembling the size and scope of a Fox News, CNN, or BBC. By some estimates, there are now almost 500 anti-vaccine misinformation websites on the Internet, all amplified on Facebook and other forms of social media, as well as e-commerce platforms [22]. The largest e-commerce platform of them all, Amazon, is now the most active promoter of fake anti-vaccine books. For example, the book about my daughter Rachel ranks among the best-selling books summarizing the benefits of vaccines and global vaccination programs. However, overall my book ranks around 25 or 30 in terms of books on the topic of vaccinations, and almost all those that place higher are essentially books that promote fake news by claiming that vaccines cause autism or various illnesses. You can do this at home: Go to Amazon books, click on "Health, Fitness, and Dieting" on the scroll down menu at the left, and then click on "Vaccinations" to see how legitimate books on vaccines are pushed behind by the fake ones. By dominating the

Internet, the anti-vaccine movement inundates parents with misinformation. Indeed, based on my experience, I conclude that it is now difficult for worried moms and dads to download accurate healthcare information about vaccines. Serious and meaningful information regarding this topic resembles a lost message in a bottle floating aimlessly in the Atlantic Ocean.

A second component of the anti-vaccine movement is its aggressive political arm. In Italy and many US states, the anti-vaccine lobby has tied itself to populist movements. For example, in states such as Texas, Oklahoma, and Colorado, it links to the libertarian sentiments of the far right wing of the Republican Party. The American "Tea Party" now justifies withholding vaccines from children by using slogans such as "medical freedom," "health freedom," or "choice." In these states, the anti-vaccine movement has spun off political action committees (PACs) to lobby state legislatures to make it easier for parents to opt their children out of school requirements for vaccines [18, 23]. PACs are especially active in the Pacific Northwest and the American Southwest, where vaccine exemption rates are high [24]. In some cases, anti-vaccine PACs raise money for political candidates. Can we trace the 2019 American measles epidemic to these political activities? Of the 14 metropolitan areas in western states and the state of Michigan where in 2018 we identified large numbers of children not receiving vaccines [24], most if not all were located in states where anti-vaccine PACs are active. Measles cases appeared in 7 of those 14 counties in 2019. In contrast, there are few if any lobbies or PACs specifically committed to vaccines.

The third component of the anti-vaccine movement is what I term "deliberate predation." It was Lena Sun of the *Washington Post* who first reported how ringleaders of the anti-vaccine movement deliberately targeted a Somali immigrant community in Minnesota to mislead them into thinking vaccines cause autism. Through contrived town hall meetings and other actions, the anti-vaccine lobby caused precipitous declines in measles vaccination coverage that resulted in a terrible epidemic in 2019, hospitalizing at least 20 people [21]. Then in 2019, a predatory group of anti-vaccine agitators again focused their activities on a specific ethnic group, in this case an Orthodox Jewish community, to cause one of America's worst measles epidemics in decades—more than 600 cases resulting in over 50 hospitalizations and 18 admissions to intensive care units [25]. Their methods included flooding the community with a pamphlet and telephone hotline containing vaccine misinformation [26], as well as hosting town hall meetings. In both Minnesota and New York, the predatory behavior of the anti-vaccine leaders manipulated local community and religious leaders to question the safety of vaccines and to make them believe that vaccines cause autism or other conditions.

Incredibly, their activities continue unopposed. In the Orthodox Jewish community, the anti-vaccine leaders used fake Holocaust imagery, including yellow stars, to compare vaccines to the Holocaust. Now, they are targeting the African American community in Harlem, New York, and comparing vaccines to the infamous Tuskegee syphilis trials.

Fighting Back

I love being a scientist and being a part of the scientific community. I take great pride in seeing young scientists give presentations at our weekly laboratory meetings and at national and international conferences. Even though I have been a scientist since 1980, when I began my MD PhD program at Rockefeller University, I still feel a special thrill when I get a paper accepted in a scientific journal or if I receive a good score on a grant application. I remain in awe of nature and feel blessed to learn something new about its secrets through the discoveries made by our team of dedicated scientists. I especially like mentoring young scientists and seeing their eyes light up when I discuss their career options and paths.

For me, science goes beyond the laboratory and the walls of academia. My dad, Eddie Hotez, was a highly pragmatic individual who served in the Pacific Theater in World War II as a 19-year-old ensign on a landing ship transport in Okinawa, Saipan, and the Philippines. He taught me the importance of giving back and doing good. I'm a true believer in the importance of basic discoveries. I devour papers on fundamental biological sciences in *PLOS Biology*, *Cell*, *Nature*, and *Science* and truly appreciate the excitement of new findings. However, my upbringing was such that I also aspired to build new therapies. I devoted my life to translating scientific discoveries into new vaccines, and I am truly grateful to Baylor College of Medicine and Texas Children's Hospital for those opportunities.

Eddie Hotez also taught me the importance of a good and noble fight. Starting in 2000, after the launch of the Millennium Development Goals, I began to champion the causes of providing access to essential NTD medicines for the world's poorest people and raising awareness about NTDs among the poor in America. These public efforts brought some successes and resulted in legislative action in the US Congress that led to tangible results in terms of improving health. Would similar efforts also work against an anti-science movement that was bringing back vaccine-preventable diseases? As a vaccine scientist and pediatrician horrified by the rise of a fake anti-vaccine movement sweeping America and depriving children of immunizations—all because of phony ideologies—I felt it was essential to step out of the laboratory to engage in battle yet again.

In 2016 and 2017, I wrote a series of articles to try to reduce the impact of the anti-vaccine movement in Texas and nationally, before writing my latest book, about Rachel. By 2018 and 2019, I started doing regular radio and TV interviews and even high-profile podcasts, including a two-hour interview with Joe Rogan. However, the anti-vaccine movement continued to gain strength and expand its media and political dimensions. In an opinion piece published in 2019 at the height of the US measles epidemic, I outlined my three-part action plan (targeted at the three components of the anti-vaccination movement described above) to stand up to the anti-vaccine movement and restore vaccine confidence in America [27].

A key component is a public policy to shut down personal belief vaccine exemptions. Relocating to Texas in 2011 taught me the impact of dangerous anti-vaccine PACs. According to the Texas Department of State Health Services, more than 60,000 children are denied access to vaccines by parents who erroneously believe that vaccines represent dangerous pharmaceuticals or cause autism. In fact, in Texas it is common to refer to autism as a form of "vaccine injury." Moreover, there are now more than 300,000 homeschooled children in the state, and we have no idea what percentage of those children do not receive vaccines. Therefore, it is likely that more than 100,000 children are currently at risk for measles and other serious infectious diseases. The number of exemptions is also high in at least 17 other states. In most of these states, there are PACs committed to keeping those exemptions open. We need to find a mechanism to counteract these dangerous political activities and halt the exemptions.

However, closing vaccine exemptions will not win hearts and minds. Therefore, in parallel we need to do something to counter the anti-vaccine misinformation campaign. In many cases, the ringleaders of the anti-vaccine movement are monetizing the Internet by selling phony autism therapies (including bleach enemas) and nutritional supplements, fake books, or advertising. Toward that goal, I recommend dismantling the anti-vaccine media empire by taking down content on Facebook, Amazon, and other social media and e-commerce sites. My stance on this has caused an aggressive reaction from the anti-vaccine community, who accuse of me of violating the First Amendment to the US Constitution or even of modern-day "book burning." I respond by pointing out the Facebook, Amazon, and other sites are private entities—not the federal government—and they have a right to select content, as does any other bookseller. In a 2019 landmark Salzburg Statement on Vaccination Acceptance, a group of vaccine experts, including myself, also reaffirmed my position [28].

Finally, I think we need to restore a more robust system of vaccine advocacy in the United States and Europe. For too long our government agencies have taken for granted the assumption that the American and European public accept vaccines as safe and lifesaving technologies. However, we are now many decades away from the 1950s and 1960s, when scientists such as Jonas Salk and Albert Sabin, the discoverers of the polio vaccines, were lauded as heroes. We need to rebuild faith and confidence in vaccines and mount an important series of public service announcements touting the efficacy of vaccines and their unparalleled safety record. Recently, the Australian government announced a $12 million campaign along these lines [29].

In the fall of 2019, the anti-vaccine movement stepped up their attacks against me and other scientists. They surrounded us at an infectious diseases of children conference held at the New York Sheraton located in Times Square, and I had to be whisked out by security into a waiting Uber. Robert F. Kennedy Jr., now one of the most vocal anti-vaccine leaders, launched a bizarre attack against me on his Instagram. He claimed that, because I received grants during the 2000s from the Gates Foundation for our hookworm vaccine and a small grant from a start-up biotech in the 1990s to develop new cancer treatments, I must be a paid "shill" for the vaccine industry, even though we receive no funding, and I receive no financial support. He also went on a paranoid rant about Jonas Salk and declared that I consider Salk a hero (actually, he got that part right). Later on his Instagram account he called me the "OG [original gangster] Villain." I worry about Kennedy's attacks because they may provoke some of his unhinged followers to further attacks or violence. Sadly, this is what it means to defend science in the twenty-first century. Defending vaccines must now become an added aspect of vaccine diplomacy.

Going Global

I am also concerned about the anti-vaccine movement in America and Europe for its potential to expand beyond the Northern Hemisphere. I believe it is only a matter of time before it goes global and spreads to Africa, Asia, and Latin America, where declines in vaccine coverage could have catastrophic consequences or even reverse global development goals [30]. So far, the evidence that this has happened is not strong, but in conversations with vaccine colleagues in some of the big low- and middle-income nations such as Bangladesh, Brazil, China, India, and Nigeria, I am starting to hear reports about how the upper classes are taking the lead from America. After all, the US exports its music and movies, so why not its vaccine misinformation?

In 2019, the WHO listed vaccine hesitancy as one of our planet's 10 leading global health threats [31]. Its concern was based on the sudden uptick of measles in North America and Europe but also on new and large measles epidemics that spread in Madagascar and the Philippines, to a point where global measles cases suddenly increased by 30% [31]. Especially worrisome was the WHO finding that some nations saw a measles resurgence just as they got close to eliminating the disease [31]. Emerging news reports indicate that the anti-vaccine movement from America had a hand in promoting vaccination declines in Samoa in 2019, which resulted in more than 80 deaths, overwhelmingly in young children. This will require a complete investigation from the Samoan government.

Still another worry is the slow progress in immunizing against cervical cancer and other vaccine-preventable diseases. The Australian government has made a commitment to advance an initiative for eliminating cervical cancer in the next decade through expanded immunization with the HPV vaccine. However, in many US states and other nations such as Japan, the rates of HPV vaccination are extremely low. In such areas, a generation of teenagers will miss out on their cancer prevention. This is because of fake assertions from the anti-vaccine lobby claiming that HPV vaccine causes autoimmune diseases, miscarriages, thromboembolic events, or even teenage suicide and depression. In so doing, the anti-vaccine movement will unnecessarily condemn women to cervical cancer and men and women to throat and other cancers. Therefore, one of the latest battlefronts is to show how the anti-vaccine movement is blocking the introduction of this and other vaccines, including annual influenza vaccinations. I'm concerned it will affect the roll-out of potential COVID-19 vaccines. The Internet and social media are already filling up with accounts claiming that the SARS CoV2 virus was engineered to sell new vaccines or that vaccine development is being accelerated to inflict dangerous vaccines on unsuspecting populations. I am now alleged to have created SARS CoV2 for this purpose or to have collaborated with Bill Gates in secret hideaway labs. The fact that we have a coronavirus vaccine program and aspire to develop new global health vaccines for COVID-19 only fuels such crazy conspiracy theories. I am also worried that a globalizing anti-vaccine lobby might affect the introduction of other urgently needed neglected disease vaccines, such as those to prevent malaria, dengue, or our vaccines for schistosomiasis, hookworm infection, and Chagas disease. In the coming decade, we will have to contend with the anti-vaccine movement as a growing and potent new force that could further promote the emergence of disease. Anti-science could become a driver as important as political unrest, poverty, urbanization, or climate change. It also represents our latest threat to global health security.

10: Implementing Vaccine Diplomacy and the Rise of COVID-19

Through vaccine diplomacy, new cholera and Ebola vaccines became powerful disease prevention tools in settings of extreme political instability or outright conflict. In DR Congo, I believe the Ebola vaccine prevented an epidemic and human catastrophe that could have easily dwarfed the one in western Africa in 2014. It potentially prevented the destabilization of much of the African continent. However, for the majority of the diseases now emerging because of conflict or political instability, internal displacements, climate change, and other modern forces, we do not yet have licensed vaccines. I consider this one of our most urgent health investment priorities, especially since we have just learned the hard lesson that vaccines can address both poverty and global security. Moreover, vaccine diplomacy offers one of our most potent yet often ignored tools to stabilize nations and effect peace.

Since the 1980s, I have had a passion and commitment to develop vaccines for the poverty-related neglected diseases, especially the 20 tropical infections also known as the NTDs. They include vaccines for leishmaniasis, Chagas disease, hookworm infection, and schistosomiasis, now recognized as the most common afflictions of people living in extreme poverty and under duress. Together, the NTDs and malaria represent some of our gravest global health threats. The most recent estimates from the Global Burden of Disease Study indicate that they kill an estimated 720,000 people annually and result in 62.3 million years lived with disability [1, 2]. Neglected diseases arise from political instability and themselves destabilize populations, owing to their long-term disabling features. Still another major global health threat is tuberculosis, perhaps representing the single leading infectious disease killer of people globally. So far, there is no licensed vaccine available for malaria, tuberculosis, or any of the NTDs caused by parasitic diseases. The only licensed NTD vaccines are for dengue and rabies.

There are formidable scientific challenges to developing new vaccines for neglected diseases. Among them is identifying suitable targets for developing a vaccine. For example, the size of the genome of a parasite such as a trypanosome, schistosome, or hookworm is roughly the size of the human genome. Parasites are complex eukaryotic organisms, and mining their genomes and distilling all the potential gene-product candidates down to a few can become a complex undertaking. Together with Dr. Kamal Rawal at Amity University in India, we are applying machine-learning technologies to help in this approach. Another hurdle is the problem of developing industrial processes to produce large quantities of the vaccine, in addition to the preclinical testing of vaccines in laboratory animals and the regulatory and safety hurdles required to successfully transition newly discovered vaccines to the clinic.

Despite such challenges, the Texas Children's Center for Vaccine Development (co-headed by Dr. Maria Elena Bottazzi and me) has developed vaccines for hookworm infection and schistosomiasis, now in clinical trials in Africa and Brazil, respectively. We also expect a new Chagas disease vaccine and a COVID-19 vaccine to enter clinical testing soon. However, the problem before us is that, after four decades of developing new NTD vaccines and advancing them to clinical testing, we have just completed the first part. Now we face an even more daunting task of reaching the final stages of clinical and product development that will lead to licensure of our vaccines.

Reaching the finish line of vaccine licensure will require us to achieve success on two fronts. First, we must develop a business model that helps us to navigate the expense and complexities of pivotal trials required by the Food and Drug Administration (FDA) or another national regulatory authority for licensure. However, there is no road map for advancing neglected disease vaccines through these steps in the nonprofit sector. We therefore urgently need innovation in the business sphere. I often explain to young people interested in a career in global health the relevance and importance of pursuing business school training as an interesting option for that path. I am convinced there is a viable (but yet undiscovered) and sustainable business model for neglected disease vaccines. In parallel, we will need a program of vaccine diplomacy in order to partner with key nations, especially those in disease-endemic countries. Our vaccines will not reach the people who need them without vaccine diplomacy.

The Coalition for Epidemic Preparedness Innovation (CEPI)

Ever since the launch of the Millennium Development Goals, global policymakers and multinational pharmaceutical companies have struggled to figure out how to develop, license, and produce at scale interventions for the poverty-related neglected diseases. Compared with other pharmaceuticals, vaccines are especially problematic because of the high cost and long time horizons required to develop these biotechnologies, including the years or even decades needed for clinical testing to ensure their safety and efficacy. For example, I began work to develop the first human hookworm vaccine as an MD PhD student at Rockefeller University in the 1980s—and only now are we entering Phase 2 clinical trials. Similarly, our schistosomiasis vaccine is now completing Phase 1 trials for safety after we began this project in the early 2000s, while our Chagas disease vaccine development program is about to enter its second decade. These long time horizons, coupled with the financial investments needed to support the science and safety studies for licensure and the absence of a promised commercial market—given that these are the diseases of the world's poorest people—translate to an elusive goal of licensing neglected disease vaccines.

Our lack of a sustainable business model for neglected disease vaccines came into a glaring spotlight during the 2014 Ebola virus epidemic in western Africa, when we were desperate to have in hand a vaccine as a critical disease-fighting tool. It was only after the Obama administration put up more than $100 million through the Biomedical Advanced Research and Development Authority [3] that the major pharmaceutical companies licensed the available technologies and began scaling up production of several different prototype vaccines for clinical testing. Ultimately, Merck & Company's rVSV-ZEBOV vaccine began showing some promise for efficacy in early 2015, but only after 11,000 people perished in Guinea, Liberia, and Sierra Leone.

Global policymakers appropriately realized that "business as usual" would not be adequate to develop the lifesaving vaccines required to combat diseases such as Ebola virus infection, and it was paramount to identify new mechanisms. The World Economic Forum at Davos offered a solution. The Gates Foundation, Wellcome Trust, and several government leaders from Norway, India, and other countries met at Davos, first in 2016 and then in 2017, with each contributing to a global fund for vaccine development and innovation. I was not present at Davos, but those deliberations and discussions created a new organization known as CEPI—the Coalition for Epidemic Preparedness Innovations.

Headquartered in Oslo and London, CEPI is a partnership between "public, private, philanthropic, and civil society organisations" [4], intended to accelerate new vaccines that the major pharmaceutical companies ordinarily would not take on. Its goal was to raise more than $2 billion to create a fund to incentivize new actors to enter into the neglected disease vaccine space. Ultimately, it raised around one-third of that amount, but it was still enough to make an ambitious beginning.

I am a supporter of CEPI and consider it to be a fellow traveler in our joint quest to develop neglected disease vaccines. Like our Texas Children's Center for Vaccine Development, the coalition is exploring innovation in both the science and business models. At the same time, I believe that the CEPI approach might benefit from some refinements as it evolves. To its credit, the Coalition collaborates closely with the WHO to identify major disease targets, including those that it would fund during first few years of operations. However, instead of focusing on the most common afflictions arising from modern Anthropocene forces, especially those of particular threat to Africa, Asia, and Latin America, CEPI mostly targets perceived pandemic threats, especially potential threats to North America, Europe, and Japan. This is not to say that new vaccines for the diseases it selected—Lassa fever, MERS coronavirus infection, and Nipah virus infection—do not affect people living in poverty; they certainly do. But there is a heavy emphasis on diseases that frighten Western leaders, now of course including COVID-19. CEPI mostly ignores the high prevalence neglected diseases that now decimate the world's poor. It means that our organization and organizations like it are still mostly on their own.

An Alternative Path

Today, most of the world's neglected disease vaccines advance through product development, clinical testing, and ultimately licensure by a national regulatory body such as the FDA through organizations known as product development partnerships (PDPs). PDPs are nonprofits, but they use industry practices to make "antipoverty" products for neglected diseases, such as drugs, diagnostics, vaccines, and new vector-control technologies [5]. Today, probably the best-known PDPs are the Drugs for Neglected Diseases Initiative and the Program for Appropriate Technology in Health, now simply known as PATH.

PATH is by far the largest of the vaccine PDPs, and through the support of the Gates Foundation, it began partnering more than a decade ago with the multinational pharmaceutical giant GlaxoSmithKline to develop the Mosquirix malaria vaccine [6]. This vaccine was first developed at Walter Reed Army Institute of Research. Now through the PATH-GlaxoSmithKline partnership, Mosquirix has completed the final Phase 3 trials, which are also known as "pivotal" trials to determine their efficacy in preventing malaria, and based on those studies the vaccine was approved for use in children by the European Medicines Agency—the FDA equivalent in Europe. The vaccine is currently being introduced into the African nations of Ghana, Kenya, and Malawi [6]. Vaccine introduction is a process that allows evaluation of the performance of the vaccine in real-life scenario situations that would ultimately resemble how it would protect children across Africa in the coming years. Similarly, GlaxoSmithKline is also advancing a new vaccine for tuberculosis in partnership with a PDP devoted to tuberculosis vaccines, known as Aeras, although recently Aeras has merged with the International AIDS Vaccine Initiative PDP [6].

Solving important public health threats like malaria can become very important for the morale of the GlaxoSmithKline employees, while simultaneously promoting company pride and team building. However, there is also a practical reason: Vaccines to prevent HIV/AIDS, tuberculosis, and malaria are "dual use" in the sense that they are intended primarily for developing country markets, which are extremely modest, but they also have some commercial potential for North America, Europe, and Japan, the traditional markets of big pharmaceutical companies. For that reason, it has been a little easier to incentivize a multinational giant like GlaxoSmithKline to work with PDPs and the Gates Foundation for advancing these neglected disease vaccines. However, for diseases such as hookworm infection, schistosomiasis, leishmaniasis, and Chagas disease, which exclusively affect people living in extreme poverty, it has been tougher to bring in multinational industry partners. Instead and so far, vaccines for these diseases have been the exclusive purview of just a few nonprofit PDPs, such as ours in Texas, PATH, the Infectious Disease Research Institute (also based in Seattle), or the International Vaccine Institute, based in Seoul, South Korea.

Partnerships for Vaccine Diplomacy

The primary mission of the vaccine PDPs is to accelerate the development of neglected disease vaccines. However, there are also potential collateral benefits. Although not generally considered a part of their original purpose, PDPs are well organized for purposes of vaccine diplomacy. For instance, as a nonprofit, it is possible to invite foreign investigators into our laboratories for purposes of vaccine development training, something not ordinarily done at multinational pharmaceutical companies or biotechs. Our Texas Children's Center for Vaccine Development regularly hosts scientists from around the world in order to instruct them in just about all aspects of vaccine development, including fermentation and scale-up process development, investigational new drug filing with the FDA, stability testing, documentation, and early stage clinical testing. In addition, because our PDP is embedded in an academic health center, we have a tradition of training doctoral students and postdoctoral scientists. On that basis, we collaborate with universities and research institutions in Brazil, Malaysia, Mexico, and Saudi Arabia for the purposes of new vaccine development training.

Our Saudi link is especially interesting because it evolved, in part, following my time as US science envoy. As it became apparent that the diseases arising from the conflicts on the Arabian Peninsula threatened the health security of the Kingdom of Saudi Arabia, I was able to work with the Saudi leadership to embark on activities linked to vaccine science diplomacy. We began by collaborating scientifically with their academic institutions, including King Saud University, with a long-range goal of tying these efforts to a nascent biotechnology industry in the country.

For Saudi Arabia, vaccine science diplomacy serves multiple purposes. First, there is an urgent need for new vaccines to protect its citizens from diseases arising in the conflict zones, such as leishmaniasis, schistosomiasis, and MERS coronavirus infection [7]. In addition, it is unlikely that the major multinational pharmaceutical companies would have a long-term interest in developing these vaccines, given their modest to almost nonexistent commercial markets. Therefore, it is critical that the Saudis take ownership of their vaccine development capacity to ensure the prevention of diseases that potentially threaten the nation's own population. In other words, the development of these vaccines is vital to the security of the country. Second, as part of Saudi Arabia's long-term economic vision, also known as Saudi Vision 2030, it is critical for the country to diversify its economy and become less dependent on oil, gas, and other fossil fuels. Biotechnology potentially helps with such economic diversification, while at the same time providing urgently needed high technology job growth. Finally, as custodian of the two holy mosques in Mecca and Medina, Saudi Arabia sees itself as the center of the Muslim world, so potentially it could one day become the major producer of halal vaccines for the Organisation of Islamic Cooperation nations.

Vaccine science diplomacy is a long-term endeavor. It typically requires one to two decades to develop and test a vaccine, in order to ensure its safety and efficacy. Therefore, embarking on programs for vaccine biotechnology requires a unique vision and understanding of the time horizons. We will simply have to wait and see whether the Kingdom of Saudi Arabia really commits to the long game. It will also be interesting to see if Saudi Arabia can learn to collaborate and form partnerships with other nations in the Middle East. Right now, vaccine development is extremely limited. Iran, through its Institut Pasteur and Razi Serum and Vaccine Institute, produces some vaccines. Would Saudi Arabia and Iran ever consider a true vaccine diplomatic collaboration? Could this help to calm Sunni-Shia tensions or surrogate conflicts in the Middle East and Central Asia? Israel also hosts a sophisticated biotechnology engine with some limited capacity for producing vaccines. Could we ever envision a Saudi-Israeli joint vaccine science initiative? Nelson Mandela's statement, "It always seems impossible until it is done," seems relevant here. I began exploring these ambitious themes as US science envoy, and maintain these aspirations in my newest role as a member of the board of governors of the US-Israel Binational Science Foundation. This organization was founded in 1972 as a bilateral agreement to promote science diplomacy and cooperation.

Outside of the Middle East, there are many other potentially promising routes to pursue vaccine science diplomacy. For example, we are exploring some exciting opportunities in India where there is substantial vaccine production capacity through organizations such as the Serum Institute of India (Pune) and Biological E (Hyderabad). Given the past successes of vaccine diplomacy between the Americans and Soviets, could the United States once again look to joint initiatives with the Russians? I believe a US-Russian initiative could become a win-win for both sides [8]. The Russians over the past few decades have underachieved in terms of their ability to produce their own vaccines, especially new ones. Yet Russia, given its expansive geography across Europe, Central Asia, and the Far East, remains highly vulnerable to the importation of new diseases, ranging from helminth infections such as opisthorchiasis to West Nile virus infection, leishmaniasis, and multidrug-resistant tuberculosis, to name a few [8]. In turn, the United States would also benefit from vaccines against these diseases, in addition to improved diplomatic ties with Russia.

Yet another example includes Latin America. Brazil, Cuba, and to some extent, Argentina and Mexico each have different levels of vaccine development capacity [9]. Indeed, our original approach for vaccine development focused on an in-depth link with Brazil's two leading public sector vaccine manufacturers, the Oswaldo Cruz Foundation (FIOCRUZ) and Instituto Butantan. For more than a decade, I was in Brazil regularly working to build vaccines jointly with these institutions—and with some successes for our vaccines to combat parasitic worm infections. One of the major reasons this program advanced was a personal relationship I developed with the two charismatic heads of FIOCRUZ and Butantan, Drs. Akira Homma and Isaias Raw, respectively. These two individuals largely built Brazil's vaccine development capacity, based on a system that ensured both quality control and assurance. However, because of age and other considerations, they eventually stepped down from leadership roles. At the same time, Brazil experienced economic downturns and significant lapses in public sector science funding under the rule of President Dilma Rousseff, as well as after her removal from office.

From this combination of factors, our decade-long productive collaborations in vaccine diplomacy with the Brazilians suffered greatly. Although we are continuing clinical trials with FIOCRUZ and the Federal University of Minas Gerais, our inability to advance a more meaningful level of vaccine diplomacy, and ultimately vaccine licensure, has been disappointing. However, I am hopeful that our collaboration may still one day be renewed and reinvigorated. In the meantime, we are now forging new Latin American links in other countries such as Panama, while in Mexico we have established a consortium of institutions focused on our therapeutic Chagas disease vaccine. Chagas disease is one of the most common neglected tropical diseases among the poor in Latin America. Through a partnership with the Carlos Slim Foundation, we are hoping to advance vaccine development in Mexico and avoid the pitfalls that slowed or halted our progress in Brazil. It has been an honor for us to work with Roberto Tapia-Conyer, who heads the Slim Foundation and his associate, Miguel Betancourt, together with several members of the Slim family, including Marco Antonio Slim Domit. The Slim family has a deep commitment to and passion for biotechnology in the region, and they are both engaged and accessible. We believe that this might represent an innovative model to advance new biotechnologies for the poorest people in Latin America.

Designing a Road Map for Anthropocene Vaccines

FIOCRUZ and Instituto Butantan belong to an innovation group known as the Developing Country Vaccine Manufacturers Network (DCVMN). DCVMN is an alliance of 50 nonprofit and for-profit vaccine manufacturers working to ensure that populations in resource-poor countries have access to high quality and affordable vaccines [10]. They meet annually to exchange best practices. In all, the member organizations of the DCVMN produce approximately 200 different vaccines and biologicals, of which the WHO prequalifies around 40 [10]. WHO prequalification helps to certify the quality of a product, allowing DCVMN manufacturers to export their vaccines to other nations [10]. A key point about the DCVMN is that it represents a collection of organizations with differing degrees of independence from the major multinational pharmaceutical organizations and, therefore, with some level of autonomy in terms of developing vaccines to prevent diseases of regional importance. One weakness of the DCVMN is that its members do not produce many novel vaccines or new vaccines for emerging infections. Instead, they often focus on reproducing established industrial practices and procedures for derivative vaccine products. However, this situation is rapidly evolving, and our PDP and some of the other PDPs are working closely with the DCVMN member organizations to build capacity for producing new vaccines.

The attraction of PDP and DCVMN partnerships is that they do not depend on the multinational pharmaceutical companies, which enter or exit this space depending on corporate leadership. Such collaborations therefore represent a potential autonomous path for vaccine diplomacy and scientific innovation. However, the product portfolio and the pipeline of new vaccines for emerging and neglected diseases arising from Anthropocene forces are so far very modest and urgently need to expand. Many of the DCVMN member organizations operate at slim margins and cannot afford the investment required to develop new vaccines, even if the diseases they prevent represent important health security threats to a country. The anticipated low returns on investment are also a barrier. For that reason, the governments of the nations hosting the DCVMN member organizations need to play a greater role in supporting public investments for these vaccines. Many of these nations, such as Argentina, Brazil, China, India, Indonesia, Mexico, Saudi Arabia, and South Africa, are actually G20 nations with large and robust economies, despite the fact that they suffer high prevalence rates of poverty-related neglected diseases and NTDs.

A potential road map for Anthropocene vaccines might proceed along the following lines. A DCVMN vaccine organization enters into a collaborative agreement with one of the vaccine PDPs and a G20 national government, possibly through a ministry of health or science and technology, to develop and produce a new vaccine for an emerging or neglected disease. That nation in turn shapes a vaccine diplomatic effort to collaborate with scientists from a different nation. For example, our Texas Children's Center for Vaccine Development could collaborate with Saudi academic and industrial institutions to develop a leishmaniasis vaccine, with support from the Saudi government or its major research organization, known as King Abdulaziz City for Science and Technology. In turn, this Saudi consortium could embark on joint collaborations with other nations in the Middle East where leishmaniasis is endemic, such as Lebanon, Turkey, or Jordan, or an even more ambitious initiative with Iran or Israel.

Still another path to innovation relies on the creation of global health technology funds from the G20 nations. Both Japan and South Korea have taken this approach, with the restriction that it involves indigenous life sciences companies and academic institutions. This is still a relatively new model, and one helped through co-financing from the Gates Foundation. I recently joined the board of directors for the South Korean fund, and it will be interesting to see how this approach evolves. Given our findings about the widespread but often hidden burden of neglected diseases among the G20 countries, I would like to see each of these nations develop similar innovation funds for incentivizing global health technologies. I also believe this represents a major contribution from the Gates Foundation for ensuring sustainable approaches to innovation.

Beyond the G20 nations, there are also several resource-poor nations that nevertheless have the scientific potential to develop or actually produce nuclear technologies [11]. They include Iran, North Korea, Pakistan, and possibly others. There remains an extraordinary opportunity for these nations to redirect their scientific prowess to the development of neglected disease vaccines. My point here is that if a nation acquires the revenue stream and scientific capability to develop nuclear technology and weapons, it almost certainly can make vaccines. I believe such redirection would represent the highest expression of vaccine science diplomacy. Again, given that the United States and the Soviets audaciously put aside their ideologies decades ago to develop and administer polio and smallpox vaccines, we should aim high in this space.

COVID-19 and Vaccine Diplomacy

Large coronavirus epidemics have become an important new threat in the twenty-first century. Previously, human coronavirus infections were considered moderately important causes of upper respiratory infections—sore throat, cough, and cold symptoms, which sometimes resulted in more serious lower respiratory tract pneumonias. Then in 2003, a devastating SARS epidemic emerged in southern China and spread to Toronto, Canada, ultimately causing more than 8,000 cases with a mortality rate of 10%. Among older individuals, the mortality approached 50%. The world was caught off guard by this new virus pathogen, the SARS CoV, and it prompted the WHO and many nations to adopt a new set of IHR (2005) and the Global Health Security Agenda. Then in 2012, MERS erupted on the Arabian Peninsula and caused serious epidemics in Saudi Arabia and South Korea. Both of these coronavirus epidemics gave us ample warning that coronaviruses are serious pandemic threats. They are not only highly transmissible respiratory virus illnesses, but in addition, they cause particular devastation in hospitals and among healthcare providers. Both SARS and MERS resulted in serious hospital infections and were associated with high attack rates among doctors and nurses. SARS even led to the death of my colleague Dr. Carlo Urbani, an Italian doctor investigating the epidemic in Vietnam for the WHO. Carlo was an important public health physician and scientist who worked on the control of parasitic helminth infections—we were fellow travelers in building the framework of the neglected tropical diseases in the early 2000s.

The SARS CoV Vaccine

Our research group is mostly focused on neglected parasitic infections, such as schistosomiasis, leishmaniasis and Chagas disease, but in 2010 we were introduced by my New York Blood Center colleague, Dr. Sara Lustigman, to a leading SARS and MERS virus research group, then led by Drs. Shibo Jiang and Lanying Du. Since the emergence of SARS, they had been working on developing a vaccine and solving a thorny problem in the field. Early vaccines against SARS composed of either the inactivated virus—similar to the Salk vaccine for polio—or modified versions of the smallpox virus expressing SARS antigens, were not very effective. In fact, they sometimes made laboratory animals sicker after they became infected with SARS CoV, a phenomenon known as immune enhancement. The exact mechanism is still not well understood, but it appears to involve cellular infiltrates in the lungs or liver following the induction of a specific type of immunity sometimes known as a Th17 response. The worry was that immune enhancement could be an important hurdle to developing a human vaccine. In fact, during the 1960s, investigators at the NIH and Children's Hospital of DC (now Children's National Medical Center) were conducting clinical trials on a vaccine to prevent an important infant respiratory virus, respiratory syncytial virus (RSV). Their RSV vaccine was an inactivated virus, and it was shown that many children actually did worse after they were exposed to the infection in the community; there may have even been two deaths in the vaccinated group [12]. This experience dampened enthusiasm for developing RSV or similar respiratory virus vaccines for decades, although now the Gates Foundation is pursuing new and innovative approaches to this important global virus pathogen.

Together, Shibo and Lanying found that if they used just a small piece of the outer spike protein that docks with the receptors found in human lungs instead of the whole virus or virus vectors expressing SARS antigens, they could get around the problem of immune enhancement. Known as the "receptor binding domain," vaccines using this piece of the protein induced protective immunity in laboratory animals, while minimizing or halting immune enhancement [13]. In 2011, we applied to the National Institutes of Health for support and were awarded a substantial grant to scale-up production of the SARS receptor binding domain antigen as a means toward developing a vaccine [14, 15]. This evolved into a partnership between the New York Blood Center, the Galveston National Laboratory, and our group at Baylor and the Texas Children's Center for Vaccine Development, ultimately resulting in the final manufacture of the antigen in collaboration with Walter Reed Army Institute of Research, where we previously had several of our other vaccines produced.

By the time our SARS vaccine was manufactured in 2016, however, there was not a lot of enthusiasm for accelerating it into clinical trials. The illness had disappeared. SARS cases were no longer reported, and we failed to persuade potential donors or industry partners that this was a vaccine that was worth the effort to stockpile for further use. This was a great source of both frustration and sadness for our group. We were enthusiastic about the performance of the vaccine in laboratory animals—it was highly protective against SARS CoV infection and appeared to be safe in terms of reducing immune enhancement—especially when the vaccine was formulated with an alum adjuvant. Alum is a common adjuvant (a substance added to a vaccine that enhances the immune response to an antigen) used worldwide, for example, in the HPV vaccine to prevent cervical and other cancers or in the combined diphtheria, pertussis, and tetanus vaccine.

Another advantage of our vaccine was that it was made as a recombinant protein vaccine expressed in yeast. This is relevant information for two reasons: First, there is precedent for a recombinant vaccine made in yeast, such as the hepatitis B vaccine, and second, it could be made at very low cost and therefore would be affordable for most low- and middle-income countries. But without further support, we had to mostly table the project. Fortunately, Dr. Bottazzi, my science partner for 20 years and now codirector of the Texas Children's Center for Vaccine Development, had the wisdom to maintain a stability-testing program for the SARS vaccine. This meant that we could determine if our vaccine remained stable in case we eventually found support to advance it into clinical testing.

The SARS CoV2 Vaccine

Fortunately, our SARS CoV vaccine has remained stable since its manufacture in 2016, and the technology used to produce it might find a new use. As a new coronavirus infection, now known as COVID-19, emerged in the city of Wuhan in central China toward the end of 2019, I immediately began to notice some similarities with SARS. Both SARS and COVID-19 likely emerged from the so-called wet markets of urban Chinese centers. During the 1990s, through an NIH grant, I worked extensively in Shanghai with the Institute of Parasitic Diseases, a branch of the Chinese CDC. Located in the old French legation of Shanghai, the institute was located just a few blocks from an outdoor "wet market," where multiple exotic animals were kept in small cages. I remember referring to it as "the killing fields" because of the way the vendors slaughtered the animals in front of the customers before placing them in plastic bags.

Unfortunately, those wet markets proved to be perfect breeding grounds for coronaviruses that circulated in bats to eventually infect exotic animals such as palm civets or pangolins, before eventually evolving rapidly to infect humans. The new coronavirus emerging from these wet markets began infecting the human population in Wuhan, and eventually caused a widespread epidemic across the urban centers of central China, infecting more than 80,000 people and causing at least 3,000 deaths. Like SARS and MERS, COVID-19 also infected large numbers of doctors, nurses, and other healthcare workers, as many as 1,700 with at least five deaths in China [16]. Later, the virus emerged in Europe, where more than 30 healthcare workers died [17]. COVID-19 then spread across cities in the United States, resulting in large numbers of deaths, and healthcare workers sickened in New York City, New Orleans, Detroit, Chicago, Houston, Miami, and Los Angeles.

In the major COVID-19 affected countries, many of the chief twenty-first-century forces discussed in this book, including war, political collapse, climate change, and anti-science, have not played a major role in promoting disease emergence. However, crowding, urbanization, and high population density are key drivers, especially in the cities of central China and in some US urban areas, particularly New York City. It has been exceptionally problematic to practice social distancing in the poor neighborhoods of cities such as New Orleans, where underlying diabetes and hypertension contribute to high case fatality rates.

As COVID-19 spread across China in January 2020, Chinese scientists began releasing reports on their studies of its causative agent. Often they used a free and open access preprint service organized by Cold Spring Harbor Laboratory on Long Island, New York, known as Biorxiv and Medrxiv. Despite what is often claimed—that the Chinese were not transparent about their COVID-19 epidemic—my personal experience with their scientists at the bench paints a different picture. Throughout January and February, I would wake up each morning to review the latest information coming out of China. It revealed that the new virus agent was closely related to the SARS CoV, and it was ultimately named SARS CoV2. SARS CoV2 exhibited approximately 80% genetic similarity to SARS CoV and bound to the same host cell receptor in the lungs. It quickly became apparent that the two viruses were similar enough that it was possible our SARS CoV vaccine manufacturing processes could be repurposed to produce a similar vaccine, but one specific for SARS CoV2. Our team of scientists worked long hours and often seven days a week throughout the spring and summer of 2020 to make this happen.

These findings ignited a flurry of activity, leading us to form a partnership with PATH, the largest of the nonprofit product development partnerships, and one with extensive experience in accelerating new vaccines for global health. For example, PATH helped to lead the development and introduction of the vaccine to prevent meningococcal A infection (together with the Serum Institute of India) and the malaria vaccine (together with GlaxoSmithKline) for Africa. Because our SARS CoV2 vaccine is produced in yeast at low cost, we feel it has the potential to serve as the first COVID-19 vaccine specifically designed for low- and middle-income nations. In contrast, many of the vaccines now being accelerated for the United States utilize innovative but potentially expensive platform technologies. We developed the receptor binding domain specific to the SARS CoV2 and, in August of 2020, licensed it to Biological E, a large DCVMN vaccine manufacturer in India. Initially, Iran was the largest middle-income nation with high numbers of COVID-19 cases, but we expect nations such as Ecuador, India, the Philippines, and some African nations to eventually become important endemic foci of COVID-19. We are already learning about the devastation of COVID-19 in some of the large and crowded urban areas of these nations. As the SARS CoV2 enters the Southern Hemisphere and "Global South," we hope our vaccine becomes the first specifically designed as safe and affordable for the world's poorest people. It could become the first new major vaccine designed for the new megacities in these regions.

Still another unexpected twist on the emergence of COVID-19 has been my role in talking to the nation about the scientific aspects of the disease and a global race to develop vaccines for this condition. Each day in the spring and summer of 2020, I have been speaking on the major cable networks—CNN, Fox News, or MSNBC—about the latest scientific advances connected with this disease and tracing the evolution of the COVID-19 pandemic, especially in the United States. I believe I might be one of the few talking heads going on all three news channels, given that each one maintains strong political views both for and against President Trump and how the White House is handling matters. I have worked hard to maintain a presence on these three channels as a way to explain to the nation that science can and should transcend politics and that, to prove it, I should feel equally comfortable having discussions with important anchors from opposite sides of the political spectrum. So far, I have been successful in threading that needle, and it has been a source of deep personal satisfaction that I can go from highly visible anchors, such as Brianna Keilar, Wolf Blitzer, Alisyn Camerota, and John Berman (CNN) to Nicolle Wallace (MSNBC) to Sandra Smith, Bill Hemmer, or Harris Faulkner (Fox News) during the day, or from Anderson Cooper (CNN) to Chris Hayes and Lawrence O'Donnell (MSNBC) at night. Dr. Sanjay Gupta has been especially nice, giving me words of encouragement and advice. And people and organizations like Joe Rogan, Dr. Oz, Alyssa Milano, and *The Daily Show* have been amazing in giving me an opportunity to speak to nationwide audiences.

This visibility has its challenges, but I believe it is a vital activity. My premise is that an important enabling factor in the rise of anti-science movements is the low profile of scientists, especially in America. Speaking to the nation about science in a time of crisis may prove to have benefits that go beyond COVID-19. Yet another challenge has been the conversion of our bedroom into a poor man's television studio. My wife, Ann, has made this her mission, and I'm eternally grateful. She has ascended ladders to tape t-shirts over the windows to block out glaring sunlight and ingeniously constructed a makeshift scholarly background in the corner to provide me with an air of respectability as I speak on Zoom, Skype, or Cisco in my new uniform of a button-down shirt, bowtie, and sweatpants. Meanwhile, while I'm speaking, the cat is sound asleep on the other side of my laptop, and Ann goes downstairs to watch me on television and then provide me with a detailed critique, or to keep me awake with coffee or green tea. Since we are social distancing in the house with our special needs daughter, Rachel, Ann sometimes has to block the doorway to prevent her from entering our bedroom while I'm interviewing. In between television and other interviews, I am on continuous teleconferences with Dr. Bottazzi and our amazing team of laboratory scientists working day and night to advance our COVID-19 vaccines into clinical trials, or with potential donors raising funds to support the product and clinical development of the vaccine. Or I am writing papers on our vaccine and working to keep up with the fast-paced scientific literature. The bottom line is that there is no sleep in this time of COVID-19.

Avoiding a Vaccine Diplomacy Failure

As we work furiously to develop a low-cost COVID-19 vaccine for global health, the US has unveiled its "Operation Warp Speed" to accelerate vaccines using cutting-edge mRNA, DNA, and adenovirus-vectored technologies. My worry is that such vaccines might become too expensive for the world's low- and middle-income countries or that they will not be made available in some cases. Now, Gavi has established a new COVAX facility in collaboration with the WHO and CEPI to address this situation. In the meantime, we hope that our low-cost vaccine will become a useful tool to fight the COVID-19 pandemic. Because it utilizes the same microbial fermentation technology in yeast that was used to produce the hepatitis B vaccine in countries such as Brazil, Cuba, India, and Indonesia, the hope is that our vaccine might be one of the first specifically developed for the world's poorest people.

11:
The Broken Obelisk

Vaccine diplomacy offers an approach and way of thinking to solve a pressing problem. We urgently need new vaccines to combat the return or emergence of diseases arising in the Anthropocene. Beyond the actual vaccines themselves, vaccine diplomacy offers a tactic to address urgencies that grow out of the framework for international scientific cooperation and collaboration, especially in times of conflict or political instability. It was a proven strategy for building bridges during the Cold War, and it is still relevant as an innovative means to waging peace. Since the 1950s, vaccine diplomacy has produced some substantial public health victories, including the eradication of smallpox, the near-elimination of polio, and the prevention of a public health catastrophe related to Ebola virus infection. It has also led to the development or refinement of vaccines to prevent these diseases. We will need to invoke vaccine diplomacy for combating COVID-19.

The greater issue is that in each of these four cases—smallpox, polio, Ebola, and COVID-19—the global health community had to respond to a crisis and scramble to rapidly develop, test, license, and distribute these vaccines. Could we also implement an anticipatory system in which nations prioritize vaccine diplomacy and routinely employ it to improve international relations? The Global Health Security Agenda does not currently emphasize vaccine development, although new organizations like CEPI and start-up innovation funds from the Japanese and South Korean governments represent promising steps toward global vaccine diplomacy. I am an enthusiastic champion of their efforts. However, I also believe that an opportunity exists for a more comprehensive effort to tackle the world's most prevalent poverty-related neglected diseases while simultaneously expanding international scientific cooperation as a core element.

The answer might be found somewhere in the G20. Today, the G20 nations, comprising 19 countries and the European Union, represent almost 90% of the global economy, yet they also harbor most of the world's poverty-neglected diseases, including the NTDs. My finding that poverty-related neglected diseases are now widespread among the G20 nations has relevance. While the G20 originally formed as an organization of central bank governors and finance ministers, starting in 2008 following the global recession, its remit expanded, and the G20 leaders began meeting annually, most recently in 2019 in Osaka, Japan. The objective of the annual G20 summit turned to reaching consensus on areas of mutual interest and urgent priorities, and to producing official declarations, many of which now extend beyond financial cooperation [1]. For instance, in 2019, the meeting focused on climate change, artificial intelligence, and women's empowerment [1].

I would like to see vaccine diplomacy shaped as a priority for a future G20 summit, with a focus on some key areas. First would be a commitment from each of the G20 nations to confront neglected diseases. Based on my analysis of data released both by the WHO and the Global Burden of Disease Study, curing and preventing disease among the poor and vulnerable populations living amid wealth in the G20 has the potential to eliminate more than two-thirds of the world's neglected diseases [2].

Research and development is central to addressing global neglected diseases. Along those lines, every member country of the G20 should agree to establish national innovation funds to promote the development of neglected disease vaccines. The national innovation funds created by the Japanese and South Korean governments (in collaboration with their indigenous life sciences industries and the Gates Foundation) represent interesting templates, but we need greater participation and similar initiatives from all of the G20 nations, especially the large, middle-income BRICS countries (Brazil, Russia, India, China, and South Africa).

Creating innovation funds for neglected disease vaccines would constitute an important step toward vaccine diplomacy, but such financial mechanisms alone are not a panacea. We still need a better system in place to promote international scientific cooperation. By harnessing the resources of the Developing Country Vaccine Manufacturers Network, the remaining PDPs committed to vaccines, and the scientific leadership of the G20 nations in collaboration with leading and specialized UN agencies, we can create an improved and efficient vaccine diplomacy ecosystem. It will take time, thought, and planning to determine exactly how this should be structured, but convening the scientific leadership of the G20 would create some important paradigm shifts and elevate vaccine diplomacy to the stature it deserves, given its modern history. I previously worked with the WHO Special Program for Research and Training on Tropical Disease Research and the World Intellectual Property Organization to move in this direction, but the development of such a cooperative system requires a higher-level champion who understands both the opportunity and urgency of vaccine diplomacy. Still another dimension is the human rights aspect of vaccine diplomacy and access to innovation and immunization—prospects that might provide a vital hook to entice such a champion and leader.

Access to Vaccines

The human rights dimension of vaccine diplomacy deserves more attention. While defending vaccines and combating the anti-vaccine movement in America, I sought the help of New York University's Professor Arthur Caplan, one of America's leading bioethicists, in order to resolve a unique dilemma regarding vaccine access. Beginning in 2015 in several conservative US states such as Texas and others in the American West, the anti-vaccine movement started to link with the far right wing of the Republican Party, also known as the Tea Party. The modern Tea Party movement began in 2009, partly in response to a liberal Obama administration, and focused its attention on conservative activism in the service of decreased government intervention and interference in American life. It took its name from the 1773 Boston Tea Party, a protest against British taxation of the commodity, in which American colonists raided three British ships in Boston Harbor and then tossed their cargo of tea overboard.

One outcome of the anti-vaccine movement's pivot to the right in 2015 was to establish in Texas a PAC that supports campaigns against vaccines. Included among its rallying cries were new slogans such as "medical freedom," "health freedom," or "choice," meaning it should be up to parents to decide what's best for their children, and if they believe that vaccines cause autism or are somehow dangerous, then it is ultimately within their purview to not vaccinate. I appealed to Art Caplan because I felt something was amiss. It occurred to me that in the guise of "freedom" and "choice," 60,000 or more children in Texas were now in fact denied a fundamental right to immunization against serious or deadly diseases. As a result, the state of Texas and ultimately many other conservative states were vulnerable to large epidemics of measles and other vaccine-preventable diseases.

Together, Caplan and I wrote an article demonstrating that vaccination is indeed a fundamental right of childhood. We argued that "[u]nder the banner of terms such as 'medical freedom' or 'choice,' parents are, sadly, subjugating the rights of their children to be protected against disease out of misinformation and misplaced fear" [3]. To put this another way, we stated that "children are being unnecessarily placed in harm's way due to failure to heed established scientific evidence in lieu of ideological or personal beliefs" [3]. Just as children have a right to be strapped into a car seat or to wear a safety belt in a moving vehicle, we explained that they are also entitled to access to vaccinations.

As one might imagine, the anti-vaccine lobby in Texas was not happy with our paper, nor my outspoken opinions regarding vaccinations as human rights of children. However, I felt there was a precedent for my views. In 1989, the UN General Assembly introduced a landmark treaty known as the UN Convention on the Rights of the Child, which among other things recognizes the right of children to health and primary care. Caplan and I maintained that access to vaccines and vaccination programs represents a key element of this ethos [3]. Sadly, the United States today is the only nation that has not signed and ratified the convention.

Global Health, Science, and Human Rights

Vaccines now have a record of accomplishment as the most powerful biotechnologies ever invented. They also have an extended role related to the larger global health framework of universal health coverage. Lawrence Gostin at Georgetown University, WHO Director-General Dr. Tedros A. Ghebreyesus, and their colleagues, point out that the UN Declaration of Human Rights also applies to global public health policies and the concept of universal health coverage, now a central tenet of the WHO's approach to strengthening health systems [4].

As a forerunner to universal health coverage in the years immediately following the 2000 Millennium Development Goals, I championed access to essential medicines for NTDs as a human right and remember with fondness my discussions with Paul Hunt, then serving as UN special rapporteur on the right to health [5]. To date, more than one billion people annually can now access low-cost or donated medicines for intestinal parasitic infections, schistosomiasis, lymphatic filariasis, and trachoma. In 2019, we published additional evidence showing how these medicines are reducing or even eliminating some NTDs but are also delivering important collateral benefits in terms of overall reductions in child mortality [6] and strengthening the sexual and reproductive health of women [7].

Equally important to the concept of universal health coverage is the fundamental right of access to scientific innovation. I believe that the world's poor have a human right to research and development for neglected disease vaccines. Access to innovation and vaccine development ranks with access to universal health coverage. In 2019, I wrote the following: "Closing the access to innovation and translational medicine gaps for some of the world's most disenfranchised peoples—aboriginal populations and the poor living amid wealth—remains one of the great science and technology challenges in this relatively new century" [8]. Vaccine diplomacy provides that framework.

In a 2018 article, Jessica Wyndham, from the American Association for the Advancement of Science, and Margaret Weigers Vitullo, from the American Sociological Association, made a larger case for science in general as a human right [9]. They argue that, even beyond vaccines and other "material products of science and technology," human rights in the sciences extend to benefiting from scientific knowledge and information as a means to shape evidence-based policies [9]. They show how science can strengthen and empower individuals and communities, and point out that 70 years ago the UN Universal Declaration of Human Rights included the right to "share in scientific advancement and its benefits" [9].

Vaccine Ambassadors

The future of vaccine diplomacy will depend on scientists with a unique skills set. They must understand the science of vaccinology but also the essential elements of diplomacy and foreign relations. They must be committed to UN human rights declarations. Ideally, a vaccine ambassador can also understand with some depth and breadth the modern Anthropocene forces, including conflict, human migrations, urbanization, climate change, and anti-science.

Currently, interdisciplinary training, such as bridging the biomedical and social sciences, is not a strength commonly found at most universities, and there are not huge career opportunities available for those adept at vaccine diplomacy. Moreover, from the time I have spent in universities, US embassies abroad, and the US State Department, my impression is that science policy and science diplomacy are probably the least developed aspects of public policy and political science. Yet, there is no question that one of America's greatest strengths is its research universities and institutes.

Wherever I have traveled as US science envoy, the high percentage of science and health ministers (or their staff) who have trained in the Unites States has left a lasting impression. An especially large number received training at our so-called land-grant universities, meaning publicly funded US institutions of higher education, such Iowa State or Purdue Universities, which under the Morrill Acts of 1862 and 1890 were created for training in "practical" subjects such as agriculture and engineering.

Today, many of these land-grant institutions are also research powerhouses and represent national treasures. I also believe these institutions could one day become powerful tools of science diplomacy. During my travels, I found many leaders across the world who deeply admire America because of its research and training capacity and especially its research universities. We can and should do a much a better job of promoting this important aspect of American life.

The Broken Obelisk

Each evening when I am in Houston, I usually take a walk with my wife, Ann, and talk about the day's events. We live in the very interesting Montrose neighborhood, which is also the home of the renowned Menil Collection, a 30-acre "neighborhood of art" [10]. One of the very inspirational Menil pieces for me is an outdoor sculpture, known as the *Broken Obelisk*. Barnett Newman during the 1960s designed the sculpture, which sits next to a reflecting pond across from the even more famous Rothko Chapel. It is a massive piece of rust-colored steel weighing more than two tons. The exact meaning of *Broken Obelisk* is often debated and a bit elusive. Some art critics claim it somehow reflects the pyramids of ancient Egypt; to others it is an upside-down broken version of the Washington Monument, or perhaps both, but to me it represents both a fragmented world in need of repair and the promise of vaccine diplomacy.

Globally, vaccine diplomacy highlights science innovation as an international treasure. It represents one of our most noble pursuits—science for the benefit of humankind—and a vision for hope and better world. Over the past 70 years we have seen how vaccine diplomacy can work, and with stunning results. It has allowed us to wipe out three of our greatest plagues, while promoting peace and scientific cooperation. Despite such

Barnett Newman's *Broken Obelisk* in Houston, Texas.
By Wikimedia Commons user Barnett Newman, https://creativecommons.org/licenses/by/3.0/deed.en.

successes, both science and scientists are too often pushed to the margins of international affairs. This is a missed opportunity. Now we need to elevate the role of science and expand vaccine diplomacy as a central element of the alliance between nations.

Literature Cited

1: A New Post-2015 Urgency

1. Hotez PJ (2013) Forgotten People, Forgotten Diseases: The Neglected Tropical Diseases and Their Impact on Global Health and Development. Washington, DC: ASM Press.
2. Hotez PJ (2016) Blue Marble Health: An Innovative Plan to Fight Diseases of Poverty amid Wealth. Baltimore, MD: Johns Hopkins University Press.
3. World Health Organization (2019) 20 million children miss out on lifesaving measles, diphtheria and tetanus vaccines in 2018. WHO Newsroom, https://www.who.int/news-room/detail/15-07-2019-20-million-children-miss-out-on-lifesaving-measles-diphtheria-and-tetanus-vaccines-in-2018, accessed July 22, 2019.
4. Hotez PJ (2018) Vaccines Did Not Cause Rachel's Autism: My Life as a Vaccine Scientist, Pediatrician, and Autism Dad. Baltimore, MD: Johns Hopkins University Press.

2: A Cold War Legacy

1. Hampton L (2009) Albert Sabin and the coalition to eliminate polio from the Americas. Am J Public Health, January, 99(1): 34–44. doi: 10.2105/AJPH.2007.117952.
2. Hotez PJ (2014) "Vaccine diplomacy": Historical perspectives and future directions. PLOS Negl Trop Dis 8(6): e2808. doi: 10.1371/journal.pntd.0002808.

3. World Health Organization (n.d.) Global health histories. https://www.who.int/global_health_histories/background/en, accessed December 22, 2019.
4. LeDuc JW, Barry MA (2004) SARS, the first pandemic of the 21st century. Emerg Infect Dis, November, 10(11): e26. doi: 10.3201/eid1011.040797_02.
5. Hotez PJ (2017) Russian-United States vaccine science diplomacy: Preserving the legacy. PLOS Negl Trop Dis 11(5): e0005320. doi: 10.1371/journal.pntd.0005320.
6. Hotez PJ (2015) Vaccine science diplomacy: Expanding capacity to prevent emerging and neglected tropical diseases arising from Islamic State (IS)–held territories. PLOS Negl Trop Dis 9(9): e0003852. doi: 10.1371/journal.pntd.0003852.
7. Esparza J, Nitsche A, Damaso CR (2018) Beyond the myths: Novel findings for old paradigms in the history of the smallpox vaccine. PLOS Pathog 14(7): e1007082. https://doi.org/10.1371/journal.ppat.1007082.
8. Horstmann DM (1991) The Sabin live poliovirus vaccination trials in the USSR, 1959. Yale J Biol Med 64(5): 499–512.
9. Swanson W (2012) Birth of a cold war vaccine. Sci Am 306(4): 66–69.
10. World Health Organization (2019) Health as a bridge for peace: Humanitarian cease-fires project (HCFP). https://www.who.int/hac/techguidance/hbp/cease_fires/en, accessed April 19, 2019.

3: Vaccine Science Envoy

1. Global Polio Eradication Initiative (n.d.) Our mission. http://polioeradication.org/who-we-are/our-mission; Global Polio Eradication Initiative (n.d.) History of polio. http://polioeradication.org/polio-today/history-of-polio, both accessed April 27, 2019.
2. Rotary International (2019) Who we are. https://www.rotary.org/en/about-rotary, accessed April 27, 2019.
3. World Health Organization (2019) What is vaccine-derived polio? https://www.who.int/features/qa/64/en, accessed April 27, 2019.
4. Henderson DA (2009) Smallpox: The Death of a Disease. Amherst, NY: Prometheus Books.
5. Macaskill W (2017) The best person who ever lived is an unknown Ukrainian man. BoingBoing, July 30. https://boingboing.net/2015/07/30/the-best-person-who-ever-lived.html.

6. Foege WH (2011) House on Fire: The Fight to Eradicate Smallpox. Berkeley: University of California Press.
7. World Health Organization (n.d.) Smallpox. https://www.who.int/csr/disease/smallpox/en/, accessed April 28, 2019.
8. Hotez PJ (2017) Russian–United States vaccine science diplomacy: Preserving the legacy. PLOS Negl Trop Dis 11(5): e0005320. doi: 10.1371/journal.pntd.0005320.
9. Hotez PJ, Kassem M (2016) Egypt: Its artists, intellectuals, and neglected tropical diseases. PLOS Negl Trop Dis 10(12): e0005072.
10. White House (2009) The president's speech in Cairo: A new beginning. https://obamawhitehouse.archives.gov/issues/foreign-policy/presidents-speech-cairo-a-new-beginning, accessed May 4, 2019.
11. Hotez PJ (2009) The neglected tropical diseases and their devastating health and economic impact on the member nations of the Organisation of the Islamic Conference. PLOS Negl Trop Dis 3(10): e539. https://doi.org/10.1371/journal.pntd.0000539.
12. Kokomo Perspective (2009) Obama administration adopts Lugar science envoy program. November 3. http://kokomoperspective.com/news/obama-administration-adopts-lugar-science-envoy-program/article_335c05be-c8bd-11de-a130-001cc4c002e0.html, accessed December 31, 2019.
13. US Department of State (n.d.) US Science Envoy Program. https://2009-2017.state.gov/e/oes/stc/scienceenvoy/index.htm, accessed December 31, 2019.
14. Feuer S, Pollock D (2017) Terrorism in Europe: The Moroccan connection. Washington Institute for Near East Policy. https://www.washingtoninstitute.org/policy-analysis/view/terrorism-in-europe-the-moroccan-connection, accessed May 4, 2019.
15. Counter Extremism Project (2019) Tunisia: Extremism and counter-extremism https://www.counterextremism.com/countries/tunisia, accessed May 4, 2019.
16. Hotez PJ (2018) Modern Sunni-Shia conflicts and their neglected tropical diseases. PLOS Negl Trop Dis 12(2): e0006008.

4: Battling Diseases of the Anthropocene

1. El Safadi D, Merhabi S, Rafei R, Mallat H, Hamze M, Acosta-Serrano A (2019) Cutaneous leishmaniasis in north Lebanon: Re-emergence of an important neglected tropical disease. Trans R Soc

Trop Med Hyg, August 1, 113(8): 471–476. doi: 10.1093/trstmh/trz030.

2. World Economic Forum (2019) How the Middle East is suffering on the front lines of climate change. https://www.weforum.org/agenda/2019/04/middle-east-front-lines-climate-change-mena, accessed May 5, 2019.
3. Stromberg J (2013) What is the Anthropocene and are we in it? Smithsonian Magazine. https://www.smithsonianmag.com/science-nature/what-is-the-anthropocene-and-are-we-in-it-164801414, accessed May 5, 2019.
4. Waters CN, Zalasiewicz J, Summerhayes C, Barnofsky AD, Poirier C, et al. (2016) The Anthropocene is functionally and stratigraphically distinct from the Holocene. Science 351 (6269): 137.
5. Hotez PJ (2016) Neglected tropical diseases in the Anthropocene: The cases of Zika, Ebola, and other infections. PLOS Negl Trop Dis 10(4): e0004648. https://doi.org/10.1371/journal.pntd.0004648.
6. Whitmee S, Haines A, Beyrer C, Boltz F, Capon AG, de Souza Dias BF, Ezeh A, Frumkin H, Gong P, Head P, Horton R, Mace GM, Marten R, Myers SS, Nishtar S, Osofsky SA, Pattanayak SK, Pongsiri MJ, Romanelli C, Soucat A, Vega J, Yach D (2015) Safeguarding human health in the Anthropocene epoch: Report of the Rockefeller Foundation–Lancet Commission on planetary health. Lancet 386 (10007): 1973–2028. doi: 10.1016/S0140-6736(15)60901-1.
7. Cemma M (2017) What's the difference? Planetary health explained. Global Health Now, September 28. https://www.globalhealthnow.org/2017-09/whats-difference-planetary-health-explained, accessed July 25, 2019.
8. Du RY, Stanaway JD, Hotez PJ (2018) Could violent conflict derail the London Declaration on NTDs? PLOS Negl Trop Dis 12(4): e0006136. https://doi.org/10.1371/journal.pntd.0006136.
9. Beyrer C, Villar JC, Suwanvanichkij V, Singh S, Baral SD, Mills EJ (2007) Neglected diseases, civil conflicts, and the right to health. Lancet 370(9587): 619–27.
10. NASA (2019) Climate change: How do we know? https://climate.nasa.gov/evidence, accessed December 31, 2019.
11. Watts J, Hunt E (2018) Halfway to boiling: The city at 50C. Guardian, August 13. https://www.theguardian.com/cities/2018/aug/13/halfway-boiling-city-50c, accessed December 31, 2019.

12. Saha S (2019) How climate change could exacerbate conflict in the Middle East. Atlantic Council, May 14. https://www.atlanticcoun cil.org/blogs/menasource/how-climate-change-could-exacerbate -conflict-in-the-middle-east, accessed December 31, 2019.
13. Shepard D (2019) Global warming: Severe consequences for Africa. United Nations, Africa Renewal. https://www.un.org/africarenewal /magazine/december-2018-march-2019/global-warming-severe -consequences-africa, accessed December 31, 2019.
14. Barrett O-L (2019) Venezuela: Drought, mismanagement and political instability. Center for Climate Change and Security, https:// climateandsecurity.org/2019/02/07/drought-mismanagement -and-political-instability-in-venezuela, accessed December 31, 2019.
15. United Nations Sustainable Development Goals (2016) Report: Inequalities exacerbate climate impacts on poor. http://www.un .org/sustainabledevelopment/blog/2016/10/report-inequalities -exacerbate-climate-impacts-on-poor, accessed December 31, 2019.
16. Blum AJ, Hotez PJ (2018) Global "worming": Climate change and its projected general impact on human helminth infections. PLOS Negl Trop Dis 12(7): e0006370. https://doi.org/10.1371/journal. pntd.0006370.
17. Boissier J, Grech-Angelini S, Webster BL, Allienne JF, Huyse T, Mas-Coma S, Toulza E, Barré-Cardi H, Rollinson D, Kincaid-Smith J, Oleaga A, Galinier R, Foata J, Rognon A, Berry A, Mouahid G, Henneron R, Moné H, Noel H, Mitta G (2016) Outbreak of urogenital schistosomiasis in Corsica (France): An epidemiological case study. Lancet Infect Dis 16(8): 971–79.
18. Scheer R, Moss D (n.d.) Mosquito-borne diseases on the uptick—thanks to global warming. Sci Am EarthTalk. https://www.scientific american.com/article/mosquito-borne-diseases-on-the-uptick -thanks-to-global-warming.
19. Messina JP, Brady OJ, Golding N, Kraemer MUG, Wint GRW, Ray SE, Pigott DM, Shearer FM, Johnson K, Earl L, Marczak LB, Shirude S, Davis Weaver N, Gilbert M, Velayudhan R, Jones P, Jaenisch T, Scott TW, Reiner RC Jr, Hay SI (2019) The current and future global distribution and population at risk of dengue. Nat Microbiol, June 10. doi: 10.1038/s41564-019-0476-8.
20. Ryan SJ, Carlson CJ, Mordecai EA, Johnson LR (2019) Global expansion and redistribution of *Aedes*-borne virus transmission risk

with climate change. PLOS Negl Trop Dis 13(3): e0007213. https://doi.org/10.1371/journal.pntd.0007213.

21. Thomson MC, Muñoz ÁG, Cousin R, Shumake-Guillemot J (2018) Climate drivers of vector-borne diseases in Africa and their relevance to control programmes. Infect Dis Poverty, August 10, 7(1): 81. doi: 10.1186/s40249-018-0460-1.
22. Hotez PJ (2017) Global urbanization and the neglected tropical diseases. PLOS Negl Trop Dis 11(2): e0005308. https://doi.org/10.1371/journal.pntd.0005308.
23. Hotez PJ (2018) Human parasitology and parasitic diseases: Heading towards 2050. Adv Parasitol 100: 29–38. doi: 10.1016/bs.apar.2018.03.002.
24. Bill & Melinda Gates Foundation (2018) Goalkeepers. https://www.gatesfoundation.org/goalkeepers/report, and file:///C:/Users/hotez/Downloads/report_en.pdf, accessed December 24, 2018.
25. Hotez P (2019) DR Congo and Nigeria: New neglected tropical disease threats and solutions for the bottom 40%. PLOS Negl Trop Dis 13(8): e0007145. doi.org/10.1371/journal.pntd.0007145.
26. Rostami A, Riahi SM, Holland CV, Taghipour A, Khalili-Fomeshi M, Fakhri Y, Omrani VF, Hotez PJ, Gasser RB (2019) Seroprevalence estimates for toxocariasis in people worldwide: A systematic review and meta-analysis. PLOS Negl Trop Dis 13(12): e0007809. doi: 10.1371/journal.pntd.0007809.
27. Wilson ML, Krogstad DJ, Arinaitwe E, Arevalo-Herrera M, Chery L, Ferreira MU, Ndiaye D, Mathanga DP, Eapen A (2015) Urban malaria: Understanding its epidemiology, ecology, and transmission across seven diverse ICEMR network sites. Am J Trop Med Hyg 93(3 Suppl): 110–23. doi: 10.4269/ajtmh.14-0834.
28. Mehta P, Hotez PJ (2016) NTD and NCD co-morbidities: The example of dengue fever. PLOS Negl Trop Dis 10(8): e0004619. https://doi.org/10.1371/journal.pntd.0004619.
29. Hotez PJ, Ferris MT (2006) The antipoverty vaccines. Vaccine 24(31–32): 5787–99.
30. Hotez PJ, Fenwick A, Savioli L, Molyneux DH (2009) Rescuing the bottom billion through control of neglected tropical diseases. Lancet 373(9674): 1570–75. doi: 10.1016/S0140-6736(09)60233-6.
31. Hotez PJ (2018) Empowering Girls and Women through Hookworm Prevention. Am J Trop Med Hyg 98(5): 1211–12. doi: 10.4269/ajtmh.17-0934.

32. Hotez PJ, Engels D, Gyapong M, Ducker C, Malecela MN (2019) Female genital schistosomiasis. N Engl J Med, December 26, 381(26): 2493–95. doi: 10.1056/NEJMp1914709.
33. World Bank (n.d.) Poverty. https://data.worldbank.org/topic/poverty, accessed June 29, 2019.
34. Murray CJ, Ortblad KF, Guinovart C, Lim SS, Wolock TM, Roberts DA, et al. (2014) Global, regional, and national incidence and mortality for HIV, tuberculosis, and malaria during 1990–2013: A systematic analysis for the Global Burden of Disease Study 2013. Lancet 384(9947): 1005–70. doi: 10.1016/S0140-6736(14)60844-8.
35. Hotez PJ (2016) Blue Marble Health: An Innovative Plan to Fight Diseases of the Poor amid Wealth. Baltimore, MD: Johns Hopkins University Press.
36. Hotez PJ (2017) Tropical illness among the poorest of the rich. Zeit Online, July 3. https://www.zeit.de/wissen/gesundheit/2017-06/g20-states-tropical-diseases-poverty-wealth-english, accessed June 29, 2019.
37. Hotez PJ (2018) Minutes to midnight: Turning back the doomsday clock through neglected disease vaccine diplomacy. PLOS Negl Trop Dis 12(9): e0006676. doi.org/10.1371/journal.pntd.0006676.
38. Hotez PJ (2019) Globalists versus nationalists: Bridging the divide through blue marble health. PLOS Negl Trop Dis 13(7): e0007156. https://doi.org/10.1371/journal.pntd.0007156.

5: The Middle East Killing Fields

1. Hotez PJ (2018) Modern Sunni-Shia conflicts and their neglected tropical diseases. PLOS Negl Trop Dis 12(2): e0006008. https://doi.org/10.1371/journal.pntd.0006008.
2. Woltin KA, Sassenberg K, Albayrak N (2018) Regulatory focus, coping strategies and symptoms of anxiety and depression: A comparison between Syrian refugees in Turkey and Germany. PLOS One 13(10): e0206522. doi: 10.1371/journal.pone.0206522.
3. Fily F, Ronat JB, Malou N, Kanapathipillai R, Seguin C, Hussein N, Fakhri RM, Langendorf C (2019) Post-traumatic osteomyelitis in Middle East war-wounded civilians: Resistance to first-line antibiotics in selected bacteria over the decade 2006–2016. BMC Infect Dis 19(1): 103. doi: 10.1186/s12879-019-3741-9.
4. Ismail MB, Rafei R, Dabboussi F, Hamze M (2018) Tuberculosis,

war, and refugees: Spotlight on the Syrian humanitarian crisis. PLOS Pathog 14(6): e1007014. https://doi.org/10.1371/journal.ppat.1007014.

5. Bannazadeh Baghi H, Alinezhad F, Kuzmin I, Rupprecht CE (2018) A perspective on rabies in the Middle East—Beyond neglect. Vet Sci 5(3): 67. doi: 10.3390/vetsci5030067.
6. Bizri AR, Fares J, Musharrafieh U (2018) Infectious diseases in the era of refugees: Hepatitis A outbreak in Lebanon. Avicenna J Med 8(4): 147–52. doi: 10.4103/ajm.AJM_130_18.
7. Mbaeyi C, Ryan MJ, Smith P, Mahamud A, Farag N, Haithami S, Sharaf M, Jorba JC, Ehrhardt D (2017) Response to a large polio outbreak in a setting of conflict: Middle East, 2013–2015. MMWR Morb Mortal Wkly Rep 66(8): 227–31. doi: 10.15585/mmwr.mm6608a6.
8. Raslan R, El Sayegh S, Chams S, Chams N, Leone A, Hajj Hussein I (2017) Re-emerging vaccine-preventable diseases in war-affected peoples of the eastern Mediterranean region: An update. Front Public Health 5:283. doi: 10.3389/fpubh.2017.00283.
9. Hotez PJ (2018) The rise of leishmaniasis in the twenty-first century. Trans R Soc Trop Med Hyg 112: 421–22.
10. Bailey F, Mondragon-Shem K, Hotez P, Ruiz-Postigo JA, Al-Salem W, Acosta-Serrano Á, et al. (2017) A new perspective on cutaneous leishmaniasis: Implications for global prevalence and burden of disease estimates. PLOS Negl Trop Dis 11(8): e0005739. https://doi.org/10.1371/journal.pntd.0005739.
11. Bailey F, Mondragon-Shem K, Haines LR, Olabi A, Alorfi A, Ruiz-Postigo JA, et al. (2019) Cutaneous leishmaniasis and co-morbid major depressive disorder: A systematic review with burden estimates. PLOS Negl Trop Dis 13(2): e0007092. https://doi.org/10.1371/journal.pntd.0007092.
12. Du R, Hotez PJ, Al-Salem WS, Acosta-Serrano A (2016) Old World cutaneous leishmaniasis and refugee crises in the Middle East and North Africa. PLOS Negl Trop Dis 10(5): e0004545. https://doi.org/10.1371/journal.pntd.0004545.
13. Alawieh A, Musharrafieh U, Jaber A, Berry A, Ghosn N, Bizri AR (2014) Revisiting leishmaniasis in the time of war: The Syrian conflict and the Lebanese outbreak. Int J Infect Dis 29: 115–19. doi: 10.1016/j.ijid.2014.04.023.
14. El Bcheraoui C, Jumaan AO, Collison ML, Daoud F, Mokdad AH

(2018) Health in Yemen: Losing ground in war time. Global Health 14(1): 42. doi: 10.1186/s12992-018-0354-9.

15. Camacho A, Bouhenia M, Alyusfi R, Alkohlani A, Naji MAM, de Radiguès X, Abubakar AM, Almoalmi A, Seguin C, Sagrado MJ, Poncin M, McRae M, Musoke M, Rakesh A, Porten K, Haskew C, Atkins KE, Eggo RM, Azman AS, Broekhuijsen M, Saatcioglu MA, Pezzoli L, Quilici ML, Al-Mesbahy AR, Zagaria N, Luquero FJ (2018) Cholera epidemic in Yemen, 2016–18: An analysis of surveillance data. Lancet Glob Health 6(6): e680–e690. doi: 10.1016/S2214-109X(18)30230-4.
16. Federspiel F, Ali M (2018) The cholera outbreak in Yemen: Lessons learned and way forward. BMC Public Health, December 4, 18(1): 1338. doi: 10.1186/s12889-018-6227-6.
17. Waldor MK, Hotez PJ, Clemens JD (2010) A national cholera vaccine stockpile: A new humanitarian and diplomatic resource. N Engl J Med 363(24): 2279–82. doi: 10.1056/NEJMp1012300.
18. Anonymous (2018) Crisis-driven cholera resurgence switches focus to oral vaccine. Bull World Health Organ, July 1, 96(7): 446–47. doi: 10.2471/BLT.18.020718.
19. Lancet Global Health (2019) Yemen needs a concrete plan—now. Lancet Glob Health 7(1): e1. doi: 10.1016/S2214-109X(18)30536-9.
20. Almutairi MM, Alsalem WS, Hassanain M, Hotez PJ (2018) Hajj, Umrah, and the neglected tropical diseases. PLOS Negl Trop Dis 12(8): e0006539. https://doi.org/10.1371/journal.pntd.0006539.
21. World Economic Forum (2019) How the Middle East is suffering on the front lines of climate change. https://www.weforum.org/agenda/2019/04/middle-east-front-lines-climate-change-mena, accessed May 5, 2019.
22. Hotez PJ (2016) Southern Europe's coming plagues: Vector-borne neglected tropical diseases. PLOS Negl Trop Dis 10(6): e0004243. https://doi.org/10.1371/journal.pntd.0004243.
23. Danaei G, Farzadfar F, Kelishadi R, Rashidian A, Rouhani OM, Ahmadinia S, et al. (2019) Iran in transition. Lancet 393: 1984–2005.

6: Africa's "Un-Wars"

1. Dorrie P (2016) The wars ravaging Africa in 2016. National Interest, January 22. https://nationalinterest.org/blog/the-buzz/the-wars-ravaging-africa-2016-14993.

2. Molyneux DH, Hotez PJ, Fenwick A (2005) "Rapid-impact interventions": How a policy of integrated control for Africa's neglected tropical diseases could benefit the poor. PLOS Med 2(11): e336. https://doi.org/10.1371/journal.pmed.0020336.
3. Hotez PJ (2013) Forgotten People, Forgotten Diseases: The Neglected Tropical Diseases and Their Impact on Global Health and Development, Washington, DC: ASM Press.
4. Africa Center for Strategic Studies (2016) War and conflict in Africa. September 21. https://africacenter.org/spotlight/war-and-conflict-in-africa.
5. Gettleman J (2010) Africa's forever wars. Foreign Policy, February 11. https://foreignpolicy.com/2010/02/11/africas-forever-wars.
6. Wagner Z, Heft-Neal S, Bhutta ZA, Black RE, Burke M, Bendavid E (2018) Armed conflict and child mortality in Africa: A geospatial analysis. Lancet 392(10150): 857–65. doi: 10.1016/S0140-6736(18)31437-5.
7. Hotez PJ, Asojo OA, Adesina AM (2012) Nigeria: "Ground zero" for the high prevalence neglected tropical diseases. PLOS Negl Trop Dis 6(7): e1600. https://doi.org/10.1371/journal.pntd.0001600.
8. Higgins J, Adamu U, Adewara K, Aladeshawe A, Aregay A, Barau I, Berens A, Bolu O, Dutton N, Iduma N, Jones B, Kaplan B, Meleh S, Musa M, Wa Nganda G, Seaman V, Sud A, Vouillamoz S, Wiesen E (2019) Finding inhabited settlements and tracking vaccination progress: The application of satellite imagery analysis to guide the immunization response to confirmation of previously-undetected, ongoing endemic wild poliovirus transmission in Borno State, Nigeria. Int J Health Geogr 18(1): 11. doi: 10.1186/s12942-019-0175-y.
9. Saraki T (2017) Ending polio in Nigeria once and for all. Council on Foreign Relations, guest blog, October 26. https://www.cfr.org/blog/ending-polio-nigeria-once-and-all, accessed December 25, 2019.
10. Webster P (2017) Nigeria's polio endgame impeded by Boko Haram. CMAJ 189(25): E877–E878. doi: 10.1503/cmaj.1095433.
11. Sato R (2019) Effect of armed conflict on vaccination: Evidence from the Boko Haram insurgency in northeastern Nigeria. Conflict and Health 13: article no. 49.
12. Denue BA, Akawu CB, Kwayabura SA, Kida I (2018) Low case fatality during 2017 cholera outbreak in Borno State, north eastern Nigeria. Ann Afr Med 17(4): 203–9. doi: 10.4103/aam.aam_66_17.

13. Crisis Group (2019) 10 conflicts to watch in 2019. https://www.crisisgroup.org/global/10-conflicts-watch-2019, accessed January 1, 2020.
14. Al-Salem W, Herricks JR, Hotez PJ (2016) A review of visceral leishmaniasis during the conflict in South Sudan and the consequences for East African countries. Parasit Vectors 9: 460. doi: 10.1186/s13071-016-1743-7.
15. Abubakar A, Ruiz-Postigo JA, Pita J, Lado M, Ben-Ismail R, Argaw D, Alvar J (2014) Visceral leishmaniasis outbreak in South Sudan 2009–2012: Epidemiological assessment and impact of a multisectoral response. PLOS Negl Trop Dis 8(3): e2720. doi: 10.1371/journal.pntd.0002720.
16. Sunyoto T, Adam GK, Atia AM, Hamid Y, Babiker RA, Abdelrahman N, Vander Kelen C, Ritmeijer K, Alcoba G, den Boer M, Picado A, Boelaert M (2018) "Kala-azar is a dishonest disease": Community perspectives on access barriers to visceral leishmaniasis (kala-azar) diagnosis and care in southern Gadarif, Sudan. Am J Trop Med Hyg 98(4): 1091–1101. doi: 10.4269/ajtmh.17-0872.
17. Nackers F, Mueller YK, Salih N, Elhag MS, Elbadawi ME, Hammam O, Mumina A, Atia AA, Etard JF, Ritmeijer K, Chappuis F (2015) Determinants of visceral leishmaniasis: A case-control study in Gedaref State, Sudan. PLOS Negl Trop Dis 9(11): e0004187. doi: 10.1371/journal.pntd.0004187.
18. Hotez PJ (2013) Forgotten People, Forgotten Diseases: The Neglected Tropical Diseases and Their Impact on Global Health and Development, 2nd ed. Washington, DC: ASM Press.
19. Aksoy S, Buscher P, Lehane M, Solano P, Van Den Abbeele J (2017) Human African trypanosomiasis control: Achievements and challenges. PLOS Negl Trop Dis 11(4): e0005454. https://doi.org/10.1371/journal.pntd.0005454.
20. Centers for Disease Control and Prevention (n.d.) 2014–2016 Ebola outbreak in West Africa. https://www.cdc.gov/vhf/ebola/history/2014-2016-outbreak/index.html, accessed January 1, 2020.
21. Bausch DG, Schwarz L (2014) Outbreak of Ebola virus disease in Guinea: Where ecology meets economy. PLOS Negl Trop Dis 8(7): e3056. https://doi.org/10.1371/journal.pntd.0003056.
22. Schlein L (2019) UN strengthens measures to combat Ebola epidemic in DR Congo. https://www.voanews.com/a/un-strengthens-measures-to-combat-ebola-epidemic-in-dr-congo/4932346.html, accessed January 1, 2020.

23. Fine Maron D (2018) Why does Ebola keep showing up in the Democratic Republic of the Congo? Sci Am, May 11. https://www.scientificamerican.com/article/why-does-ebola-keep-showing-up-in-the-democratic-republic-of-the-congo, accessed January 1, 2020.
24. United Nations (2019) Amid "unprecedented combination" of epidemics, UN and partners begin cholera vaccination campaign in DR Congo. UN News, https://news.un.org/en/story/2019/05/1039211, accessed January 1, 2020.

7: The Northern Triangle and Collapse of Venezuela

1. Supporting nonprofit organizations include the Carlos Slim Foundation, the Kleberg Foundation, and the Southwest Electronic Energy Medical Research Institute.
2. Hotez PJ, Damania A, Bottazzi ME (2020) Central Latin America: Two decades of challenges in neglected tropical disease control. PLOS Negl Trop Dis 14(3): e0007962. https://doi.org/10.1371/journal.pntd.0007962.
3. Paniz-Mondolfi AE, Tami A, Grillet ME, Márquez M, Hernández-Villena J, Escalona-Rodríguez MA, Blohm GM, Mejías I, Urbina-Medina H, Rísquez A, Castro J, Carvajal A, Walter C, López MG, Schwabl P, Hernández-Castro L, Miles MA, Hotez PJ, Lednicky J, Morris JG Jr., Crainey J, Luz S, Ramírez JD, Sordillo E, Llewellyn M, Canache M, Araque M, Oletta J (2019) Resurgence of vaccine-preventable diseases in Venezuela as a regional public health threat in the Americas. Emerg Infect Dis 25(4): 625–32. doi: 10.3201/eid2504.181305.
4. Oroxom R, Glassman A (2018) Call a spade a spade: Venezuela is a public health emergency. Center for Global Development, September 21. https://www.cgdev.org/blog/call-spade-spade-venezuela-public-health-emergency, accessed January 1, 2020.
5. Grillet ME, Hernández-Villena JV, Llewellyn MS, Paniz-Mondolfi AE, Tami A, Vincenti-Gonzalez MF, Márquez M, Mogollon-Mendoza AC, Hernández-Pereira CE, Plaza-Morr JD, Blohm G, Grijalva MJ, Costales JA, Ferguson HM, Schwabl P, Hernández-Castro LE, Lamberton PHL, Streicker DG, Haydon DT, Miles MA, Acosta-Serrano A, Acquattela H, Basáñez MG, Benaim G, Colmenares LA, Conn JE, Espinoza R, Freilij H, Graterol-Gil MC, Hotez PJ, Kato H, Lednicky JA, Martinez CE, Mas-Coma S, Morris JG Jr., Navarro JC,

Ramírez JL, Rodríguez M, Urbina JA, Villegas L, Segovia MJ, Carrasco HJ, Crainey JL, Luz SLB, Moreno JD, Noya Gonzalez OO, Ramírez JD, Alarcón-de Noya B (2019) Venezuela's humanitarian crisis, resurgence of vector-borne diseases, and implications for spillover in the region. Lancet Infect Dis 19(5): e149–e161. doi: 10.1016/S1473-3099(18)30757-6.

6. Hotez PJ, Basáñez MG, Acosta-Serrano A, Grillet ME (2017) Venezuela and its rising vector-borne neglected diseases. PLOS Negl Trop Dis 11(6): e0005423. doi: 10.1371/journal.pntd.0005423.
7. Kaur H, Alberti M (2020) A boy from a remote Amazonian tribe has died, raising concerns about Covid-19's impact on indigenous people. CNN, April 20. https://www.cnn.com/2020/04/10/world/yanomami-amazon-coronavirus-brazil-trnd/index.html.
8. Hotez PJ (2014) The NTDs and vaccine diplomacy in Latin America: Opportunities for United States foreign policy. PLOS Negl Trop Dis 8(9): e2922. https://doi.org/10.1371/journal.pntd.0002922.
9. Bacon KM, Hotez PJ, Kruchten SD, Kamhawi S, Bottazzi ME, Valenzuela JG, Lee BY (2013) The potential economic value of a cutaneous leishmaniasis vaccine in seven endemic countries in the Americas. Vaccine 31(3): 480–86. doi: 10.1016/j.vaccine.2012.11.032.

8: Sorting It Out

1. Hotez PJ (2018) The rise of neglected tropical diseases in the "new Texas." PLOS Negl Trop Dis 12(1): e0005581. https://doi.org/10.1371/journal.pntd.0005581.
2. Hotez PJ (2016) Southern Europe's coming plagues: Vector-borne neglected tropical diseases. PLOS Negl Trop Dis 10(6): e0004243. https://doi.org/10.1371/journal.pntd.0004243.
3. Cloke H (2019) Heatwave "completely obliterated" the record for Europe's hottest ever June. The Conversation, July 3. https://theconversation.com/heatwave-completely-obliterated-the-record-for-europes-hottest-ever-june-119801, accessed July 24, 2019.

9: Global Health Security and the Rise in Anti-science

1. Heymann DL, Chen L, Takemi K, Fidler DP, et al. (2015) Global health security: The wider lessons from the west African Ebola virus disease epidemic. Lancet 385(9980): 1884–901.

2. World Health Organization (n.d.) Constitution. https://www.who.int/about/who-we-are/constitution, accessed May 6, 2020.
3. World Health Organization (n.d.) About IHR. https://www.who.int/ihr/about/en, accessed December 26, 2019.
4. World Health Organization (n.d.) Health security. https://www.who.int/health-security/en, accessed December 26, 2019.
5. Osterholm MT (2017) Global health security—An unfinished journey. Emerg Infect Dis 23(Suppl 1): S225-7.
6. Global Health Security Agenda (2019) About the GHSA. https://www.ghsagenda.org/about, accessed December 26, 2019.
7. Global Health Security Agenda (2019) Joining the GHSA. https://ghsagenda.org/home/joining-the-ghsa/, accessed December 26, 2019.
8. Waldor MK, Hotez PJ, Clemens JD (2010) A national cholera vaccine stockpile—A new humanitarian and diplomatic resource. N Engl J Med 363(24): 2279–82. doi: 10.1056/NEJMp1012300.
9. World Health Organization (2019) International Coordinating Group (ICG) on vaccine provision for cholera. https://www.who.int/csr/disease/icg/cholera/en, accessed December 26, 2019.
10. World Health Organization, Regional Office for the Eastern Mediterranean Region (2019) Fighting the world's largest cholera outbreak: Oral cholera vaccination campaign begins in Yemen. http://www.emro.who.int/yem/yemen-news/oral-cholera-vaccination-campaign-in-yemen-begins.html, accessed December 26, 2019.
11. World Health Organization (2019) The global task force on cholera control. https://www.who.int/cholera/task_force/en, accessed December 26, 2019.
12. US Department of Health and Human Services (2017) HHS accelerates development of first Ebola vaccines and drugs. https://www.hhs.gov/about/news/2017/09/29/hhs-accelerates-development-first-ebola-vaccines-and-drugs.html, accessed December 26, 2019.
13. World Health Organization (2019) WHO adapts Ebola vaccination strategy in the Democratic Republic of the Congo to account for insecurity and community feedback. https://www.who.int/news-room/detail/07-05-2019-who-adapts-ebola-vaccination-strategy-in-the-democratic-republic-of-the-congo-to-account-for-insecurity-and-community-feedback, accessed December 26, 2019.
14. World Health Organization (2019) Preliminary results on the efficacy of rVSV-ZEBOV-GP Ebola vaccine using the ring vaccination

strategy in the control of an Ebola outbreak in the Democratic Republic of the Congo: An example of integration of research into epidemic response. https://www.who.int/csr/resources/publica tions/ebola/ebola-ring-vaccination-results-12-april-2019.pdf, accessed December 26, 2019.

15. World Health Organization (2019) Vaccination in humanitarian emergencies. https://www.who.int/immunization/programmes _systems/policies_strategies/vaccination_humanitarian_emergen cies/en, accessed December 26, 2019.
16. Thornton J (2019) Measles cases tripled from 2017 to 2018. BMJ 364: l634 https://www.bmj.com/content/364/bmj.l634.full, accessed July 24, 2019.
17. Centers for Disease Control and Prevention (n.d.) Measles cases and outbreaks. https://www.cdc.gov/measles/cases-outbreaks.html, accessed July 24, 2019.
18. Hotez PJ (2016) Texas and its measles epidemics. PLOS Med 13(10): e1002153. https://doi.org/10.1371/journal.pmed.1002153.
19. Olive JK, Hotez PJ, Damania A, Nolan MS (2018) The state of the antivaccine movement in the United States: A focused examination of nonmedical exemptions in states and counties. PLOS Med 15(6): e1002578. https://doi.org/10.1371/journal.pmed.1002578.
20. Hotez P (2019) America and Europe's new normal: The return of vaccine-preventable diseases. Pediatr Res 85(7): 912–14. doi: 10.1038 /s41390-019-0354-3.
21. Hotez P (2018) Vaccines Did Not Cause Rachel's Autism: My Journey as a Vaccine Scientist, Pediatrician, and Autism Dad. Baltimore, MD: Johns Hopkins University Press.
22. Hotez P (2019) The physician-scientist: Defending vaccines and combating antiscience. J Clin Invest 130: 2169–71. doi: 10.1172 /JCI129121.
23. Caplan AL, Hotez PJ (2018) Science in the fight to uphold the rights of children. PLOS Biol 16(9): e3000010. doi: 10.1371/journal.pbio .3000010.
24. Hotez PJ (2019) The counties where the anti-vaccine movement thrives in the US. The Conversation, April 30. https://theconversa tion.com/the-counties-where-the-anti-vaccine-movement-thrives -in-the-us-106036, accessed July 24, 2019.
25. Enman S (2019) One Williamsburg school "ignited" NYC's measles crisis. Brooklyn Daily Eagle, June 25. https://brooklyneagle.com

/articles/2019/06/25/williamsburg-school-ignited-nyc-measles-crisis, accessed December 27, 2019.
26. Hogan G (2019) Misinformation hotline stokes fear of vaccines in ultra-Orthodox community. Gothamist, March 12, https://gothamist.com/2019/03/12/vaccinations.php, accessed July 24, 2019.
27. Hotez P (2019) As measles cases climb, our mission is clear: Take down the three-headed anti-vax monster. Newsweek, May 9. https://www.newsweek.com/measles-anti-vaccination-anti-vaxxers-misinformation-monster-1420977, accessed July 24, 2019.
28. Ratzan SC, Bloom BR, El-Mohandes A, Fielding J, Gostin LO, Hodge JG, Hotez P, Kurth A, Larson HJ, Nurse J, Omer SB, Orenstein WA, Salmon D, Rabin K (2019) The Salzburg statement on vaccination acceptance. J Health Commun 24(5): 1–3. doi: 10.1080/10810730.2019.1622611.
29. SBS News (2019) Government to spend extra $12m on national ad blitz to counter anti-vaxxers. February 18. https://www.sbs.com.au/news/government-to-spend-extra-12m-on-national-ad-blitz-to-counter-anti-vaxxers, accessed July 24, 2019.
30. Hotez PJ (2017) Will an American-led anti-vaccine movement subvert global health? Sci Am, March 3. https://blogs.scientificamerican.com/guest-blog/will-an-american-led-anti-vaccine-movement-subvert-global-health, accessed July 24, 2019.
31. World Health Organization (2019) Ten threats to global health in 2019. https://www.who.int/emergencies/ten-threats-to-global-health-in-2019, accessed July 24, 2019.

10: Implementing Vaccine Diplomacy and the Rise of COVID-19

1. GBD 2017 Causes of Death Collaborators (2018) Global, regional, and national age-sex-specific mortality for 282 causes of death in 195 countries and territories, 1980–2017: A systematic analysis for the Global Burden of Disease Study 2017. Lancet 392(10159): 1736–88. doi: 10.1016/S0140-6736(18)32203-7.
2. GBD 2017 DALYs and HALE Collaborators (2018) Global, regional, and national disability-adjusted life-years (DALYs) for 359 diseases and injuries and healthy life expectancy (HALE) for 195 countries and territories, 1990–2017: A systematic analysis for the Global Burden of Disease Study 2017. Lancet 392(10159): 1859–1922. doi: 10.1016/S0140-6736(18)32335-3.

3. Kates J, Michaud J, Wexler A, Valentine A (2015) The U.S. response to Ebola: Status of the FY2015 emergency Ebola appropriation. Kaiser Family Foundation, Global Health Policy, December 11. https://www.kff.org/global-health-policy/issue-brief/the-u-s-response-to-ebola-status-of-the-fy2015-emergency-ebola-appropriation, accessed December 28, 2019.
4. CEPI (n.d.) Creating a world in which epidemics are no longer a threat to humanity. https://cepi.net/about/whyweexist, accessed December 28, 2019.
5. Bottazzi ME, Hotez PJ (2019) "Running the gauntlet": Formidable challenges in advancing neglected tropical diseases vaccines from development through licensure, and a "call to action." Hum Vaccin Immunother 15(10): 2235–42. doi:10.1080/21645515.2019.1629254.
6. Hotez PJ (2019) Immunizations and vaccines: A decade of successes and reversals, and a call for "vaccine diplomacy." Int Health 11(5): 331–33. doi: 10.1093/inthealth/ihz024.
7. Hotez PJ (2015) Vaccine science diplomacy: Expanding capacity to prevent emerging and neglected tropical diseases arising from Islamic State (IS)–held territories. PLOS Negl Trop Dis 9(9): e0003852. doi: 10.1371/journal.pntd.0003852.
8. Hotez PJ (2017) Russian–United States vaccine science diplomacy: Preserving the legacy. PLOS Negl Trop Dis 11(5): e0005320. doi: 10.1371/journal.pntd.0005320.
9. Hotez PJ (2014) The NTDs and vaccine diplomacy in Latin America: Opportunities for United States foreign policy. PLOS Negl Trop Dis 8(9): e2922. doi: 10.1371/journal.pntd.0002922.
10. Developing Country Vaccine Manufacturers Network (n.d.) About DVCMN. https://www.dcvmn.org, accessed December 28, 2019.
11. Hotez PJ (2018) Minutes to midnight: Turning back the doomsday clock through neglected disease vaccine diplomacy. PLOS Negl Trop Dis 12(9): e0006676. doi: 10.1371/journal.pntd.0006676.
12. Kapikian AZ, Mitchell RH, Chanock RM, Shvedoff RA, Stewart CE (1969) An epidemiologic study of altered clinical reactivity to respiratory syncytial (RS) virus infection in children previously vaccinated with an inactivated RS virus vaccine. Am J Epidemiol 89(4): 405–21.
13. Jiang S, Bottazzi ME, Du L, Lustigman S, Tseng CT, Curti E, Jones K, Zhan B, Hotez PJ (2012) Roadmap to developing a recombinant coronavirus S protein receptor-binding domain vaccine for severe

acute respiratory syndrome. Expert Rev Vaccines 11(12): 1405–13. doi: 10.1586/erv.12.126.
14. Chen WH, Chag SM, Poongavanam MV, Biter AB, Ewere EA, Rezende W, Seid CA, Hudspeth EM, Pollet J, McAtee CP, Strych U, Bottazzi ME, Hotez PJ (2017) Optimization of the production process and characterization of the yeast-expressed SARS-CoV recombinant receptor-binding domain (RBD219-N1), a SARS vaccine candidate. J Pharm Sci 106(8): 1961–70. doi: 10.1016/j.xphs.2017.04.037.
15. Chen WH, Du L, Chag SM, Ma C, Tricoche N, Tao X, Seid CA, Hudspeth EM, Lustigman S, Tseng CT, Bottazzi ME, Hotez PJ, Zhan B, Jiang S (2014) Yeast-expressed recombinant protein of the receptor-binding domain in SARS-CoV spike protein with deglycosylated forms as a SARS vaccine candidate. Hum Vaccin Immunother 10(3): 648–58.
16. Wu V, McGoogan JM (2020) Characteristics of and important lessons from the coronavirus disease 2019 (COVID-19) outbreak in China. JAMA Network, https://jamanetwork.com/journals/jama/fullarticle/2762130.
17. Minder R, Peltier E (2020) Virus knocks thousands of health workers out of action in Europe. New York Times, March 24. https://www.nytimes.com/2020/03/24/world/europe/coronavirus-europe-covid-19.html.

11: The Broken Obelisk

1. Crowley M (2019) What is the G20? New York Times, June 27. https://www.nytimes.com/2019/06/27/world/asia/what-is-the-g20.html, accessed December 29, 2019.
2. Hotez PJ (2016) Blue Marble Health: An Innovative Plan to Fight Diseases of the Poor amid Wealth. Baltimore, MD: Johns Hopkins University Press.
3. Caplan AL, Hotez PJ (2018) Science in the fight to uphold the rights of children. PLOS Biol 16(9): e3000010. https://doi.org/10.1371/journal.pbio.3000010.
4. Gostin LO, Meier B, Thomas R, Magar V, Ghebreyesus TA (2018) 70 years of human rights in global health: Drawing on a contentious past to secure a hopeful future. Lancet 392: 2731–35.

5. Hunt P (2006) The human right to the highest attainable standard of health: New opportunities and challenges. Trans R Soc Trop Med Hyg 100(7): 603–7.
6. Hotez PJ, Fenwick A, Molyneux DH (2019) Collateral benefits of preventive chemotherapy: Expanding the war on neglected tropical diseases. N Engl J Med 380(25): 2389–91. doi: 10.1056/NEJMp1900400.
7. Hotez PJ, Engels D, Gyapong M, Ducker C, Malecela MN (2019) Female genital schistosomiasis. N Engl J Med 381(26): 2493–95. doi: 10.1056/NEJMp1914709.
8. Hotez PJ (2019) Science tikkun: A framework embracing the right of access to innovation and translational medicine on a global scale. PLOS Negl Trop Dis 13(6): e0007117. https://doi.org/10.1371/journal.pntd.0007117.
9. Wyndham JM, Weigers Vitullo M (2018) Define the human right to science. Science 362(6418): 975.
10. Menil Collection (n.d.) About the Menil. https://www.menil.org/about.

About the Author

Peter J. Hotez, MD, PhD, is dean of the National School of Tropical Medicine and professor of pediatrics, molecular virology, and microbiology at Baylor College of Medicine, where he is also the codirector of the Texas Children's Center for Vaccine Development. He is also university professor of biology at Baylor University, faculty fellow at the Hagler Institute for Advanced Study at Texas A&M University, fellow in disease and poverty at the Baker Institute for Public Policy at Rice University, and senior fellow at the Scowcroft Institute of International Affairs at Texas A&M University.

Hotez is an internationally recognized physician-scientist in neglected tropical diseases vaccine development. He obtained his undergraduate degree in molecular biophysics from Yale University in 1980, followed by a PhD degree in biochemistry from Rockefeller University in 1986, and an MD from Weill Medical College of Cornell University in 1987. Hotez has authored more than 500 original scientific papers, and he is the author of four single-author books and the editor of several textbooks. He was president of the American Society of Tropical Medicine and Hygiene, and he is founding editor in chief of *PLOS Neglected Tropical Diseases*. Hotez is an elected member of the National Academy of Medicine and the American Academy of Arts and Sciences. In 2011, he was given the Abraham Horwitz Award by the Pan American Health Organization of the WHO.

In 2014–16, Hotez served in the Obama administration as US science envoy, focusing on vaccine diplomacy initiatives between the US government and countries in the Middle East and North Africa. In 2018, he was appointed to serve on the board of governors for the US-Israel Binational Science Foundation. He has served on infectious disease task forces for two consecutive Texas governors. For these efforts, he was named in 2017 by *Fortune* magazine as one of the 34 most influential people in healthcare and received the distinguished achievement award from B'nai B'rith International. In 2018, he received the Sustained Leadership Award from Research!America.

Index

Page numbers in *italics* refer to illustrations and tables.